NITROGEN ECONOMY OF FLOODED RICE SOILS

Developments in Plant and Soil Sciences

1. J. Monteith and C. Webb, eds.,
   Soil Water and Nitrogen in Mediterranean-type Environments. 1981. ISBN 90-247-2406-6
2. J.C. Brogan, ed.,
   Nitrogen Losses and Surface Run-off from Landspreading of Manures. 1981. ISBN 90-247-2471-6
3. J.D. Bewley, ed.,
   Nitrogen and Carbon Metabolism. 1981. ISBN 90-247-2472-4
4. R. Brouwer, I. Gašparíková, J. Kolek and B.C. Loughman, eds.,
   Structure and Function of Plant Roots. 1981. ISBN 90-247-2510-0
5. Y.R. Dommergues and H.G. Diem, eds.,
   Microbiology of Tropical Soils and Plant Productivity. 1982. ISBN 90-247-2624-7
6. G.P. Robertson, R. Herrera and T. Rosswall, eds.,
   Nitrogen Cycling in Ecosystems of Latin America and the Caribbean. 1982. ISBN 90-247-2719-7
7. D. Atkinson et al., eds.,
   Tree Root Systems and their Mycorrhizas. 1983. ISBN 90-247-2821-5
8. M.R. Sarić and B.C. Loughman, eds.,
   Genetic Aspects of Plant Nutrition. 1983. ISBN 90-247-2822-3
9. J.R. Freney and J.R. Simpson, eds.,
   Gaseous Loss of Nitrogen from Plant-Soil Systems. 1983. ISBN 90-247-2820-7
10. United Nations Economic Commission for Europe.
    Efficient Use of Fertilizers in Agriculture. 1983. ISBN 90-247-2866-5
11. J. Tinsley and J.F. Darbyshire, eds.,
    Biological Processes and Soil Fertility. 1984. ISBN 90-247-2902-5
12. A.D.L. Akkermans, D. Baker, K. Huss-Danell and J.D. Tjepkema, eds.,
    Frankia Symbioses. 1984. ISBN 90-247-2967-X
13. W.S. Silver and E.C. Schröder, eds.,
    Practical Application of Azolla for Rice Production. 1984. ISBN 90-247-3068-6
14. P.G.L. Vlek, ed.,
    Micronutrients in Tropical Food Crop Production. 1985. ISBN 90-247-3085-6
15. T.P. Hignett, ed.,
    Fertilizer Manual. 1985. ISBN 90-247-3122-4
16. D. Vaughan and R.E. Malcolm, eds.,
    Soil Organic Matter and Biological Activity. 1985. ISBN 90-247-3154-2
17. D. Pasternak and A. San Pietro, eds.,
    Biosalinity in Action: Bioproduction with Saline Water. 1985. ISBN 90-247-3159-3.
18. M. Lalonde, C. Camiré and J.O. Dawson, eds.,
    Frankia and Actinorhizal Plants. 1985. ISBN 90-247-3214-X
19. H. Lambers, J.J. Neeteson and I. Stulen, eds.,
    Fundamental, Ecological and Agricultural Aspects of Nitrogen Metabolism in Higher Plants. 1986.
    ISBN 90-247-3258-1
20. M.B. Jackson, ed.
    New Root Formation in Plants and Cuttings. 1986. ISBN 90-247-3260-3
21. F.A. Skinner and P. Uomala, eds.,
    Nitrogen Fixation with Non-Legumes. 1986. ISBN 90-247-3283-2
22. A. Alexander, ed.
    Foliar Fertilization. 1986. ISBN 90-247-3288-3.
23. H.G. v.d. Meer, J.C. Ryden and G.C. Ennik, eds.,
    Nitrogen Fluxes in Intensive Grassland Systems. 1986. ISBN 90-247-3309-x.
24. A.U. Mokwunye and P.L.G. Vlek, eds.,
    Management of Nitrogen and Phosporus Fertilizers in Sub-Saharan Africa. 1986.
    ISBN 90-247-3312-x
25. Y. Chen and Y. Avnimelech, eds.,
    The Role of Organic Matter in Modern Agriculture. 1986. ISBN 90-247-3360-x
26. S.K. De Datta and W.H. Patrick Jr., eds.,
    Nitrogen Economy of Flooded Rice Soils. 1986. ISBN 90-247-3361-8

Nitrogen Economy of Flooded Rice Soils

Proceedings of a symposium on the Nitrogen Economy of Flooded Rice Soils, Washington DC, 1983

Edited by

S. K. DE DATTA
Head, Dept. of Agronomy,
International Rice Research Institute,
Manila, Philippines

W. H. PATRICK JR.,
Boyd Professor,
Center for Wetland Resources,
Louisiana State University,
Louisiana, U.S.A.

1986 **MARTINUS NIJHOFF PUBLISHERS**
a member of the KLUWER ACADEMIC PUBLISHERS GROUP
DORDRECHT / BOSTON / LANCASTER

Distributors

for the United States and Canada: Kluwer Academic Publishers, 190 Old Derby Street, Hingham, MA 02043, USA
for the UK and Ireland: Kluwer Academic Publishers, MTP Press Limited, Falcon House, Queen Square, Lancaster LA1 1RN, UK
for all other countries: Kluwer Academic Publishers Group, Distribution Center, P.O. Box 322, 3300 AH Dordrecht, The Netherlands

Library of Congress Cataloging in Publication Data

```
Nitrogen economy of flooded rice soils.

   (Developments in plant and soil sciences ; v.    )
   1. Rice--Soils--Congresses.  2. Rice--Fertilizers--
Congresses.  3. Soils--Nitrogen content--Congresses.
4. Nitrogen fertilizers--Congresses.  5. Soils,
Irrigated--Congresses.  I. De Datta, Surajit K.,
1933-    .  II. Patrick, W. H. (William H.), 1925-
III. Series.
S597.R5N58  1986      633.1'8            86-8592
ISBN 90-247-3361-8
```

ISBN 90-247-3361-8
ISBN 90-247-2405-8 (series)

Copyright

© 1986 by Martinus Nijhoff Publishers, Dordrecht.

PRINTED IN THE NETHERLANDS

CONTENTS

CONTENTS

Preface

The steadily increasing cost of nitrogen fertilizer has resulted in more emphasis on basic and applied studies to improve nitrogen use efficiency in lowland rice. The efficiency of fertilizer nitrogen in farmers' fields is shockingly low – a luxury resource-scarce farmers in tropical Asia can ill afford.

We believe it is critical to quantify the basic transformation processes and develop management practices for higher N use efficiency for two reasons. They are:

1. Nitrogen fertilizer together with water management is a key factor for achieving the yield potentials of modern rices.
2. Fertilizer nitrogen prices are high and most Asian rice farmers are poor.

The International Rice Research Institute (IRRI), Philippines; International Fertilizer Development Center (IFDC), USA; Commonwealth Scientific and Industrial Research Organization (CSIRO), Australia; U.S. Universities (Louisiana, Cornell, California, Arkansas and others); and Dr Justus Leibig University in West Germany are actively engaged in individual or collaborative research that addresses basic transformation processes on N gains and losses and management practices to maximize N use efficiency in rice.

It is appropriate to update and summarize, in a double issue of *Fertilizer Research,* the 10 papers presented at the special symposium organized by the American Society of Agronomy (ASA) at the 75th Annual Meeting in Washington, D.C. in 1983. S.K. De Datta, Head of Agronomy Department, IRRI, was chairman of the International Agronomy Division of ASA (A-6) in 1982 and 1983.

The symposium was a joint effort of Division A-6 with S-4 and S-8. E.T. Craswell, former Soil Scientist, IFDC, now Research Program Coordinator, Australian Centre for International Agriculture Research (ACIAR); W.H. Patrick, Jr, Boyd Professor, Center for Wetland Resources, Louisiana State University; and S.K. De Datta, prepared the program and identified the speakers. N.C. Brady, Senior Assistant Adminstrator for Science and Technology, Agency for International Development, U.S. (USAID), and D.L. McCune, Managing Director, IFDC, chaired the two sessions.

We hope that this proceedings of the conference will be a welcome addition to our knowledge pool on nitrogen economy in lowland rice.

IRRI, Los Banos, Philippines   S.K. DE DATTA

Lousiana State University, Baton Rouge   and
La, U.S.A.   W.H. PATRICK, Jr
1985   *Editors*

1. The chemistry and biology of flooded soils in relation to the nitrogen economy in rice fields

DR BOULDIN

Agronomy Department, Cornell University, Ithaca, NY 14853, USA

Key words: fertilizer N, soil N, ^{15}N

Abstract. Response of lowland rice to sources and methods of nitrogen fertilizer application were summarized for more than 100 experiments. In about $\frac{2}{3}$ of the experiments, the yield increase per kg of fertilizer N was judged to be relatively poor with 'best split' applications of urea. Based on frequency distribution, sulfur coated urea and urea briquets or urea supergranules deep placed more often produced satisfactory yield increases than 'best split' urea, but even with these sources/methods the yield increases were judged to be relatively poor in about $\frac{1}{2}$ of the experiments. There is an enormous potential to increase rice production with no further increases in inputs of fertilizer N if we could learn to match the best method/source of fertilizer with the soil-crop management complex.

About 60% of the yields with no fertilizer N were in the range of 2 to 4 t/ha. Based on the average yield response to urea, this is equivalent to about 100 kg of urea N. It would appear worthwhile to study ways to improve utilization of 'soil nitrogen' since it is already in place on the land and apparently in fairly abundant amounts in many soils.

About 50 experiments with ^{15}N fertilizers were summarized. In almost all cases, the uptake of tagged fertilizer was less than the net increase in N in the above ground matter. In about $\frac{2}{3}$ of the experiments, the addition of fertilizer N increased soil N uptake more than 20% and in $\frac{1}{3}$ of the experiments the uptake of soil N was increased more than 40%. These results lead to much uncertainty about practical interpretation and use of ^{15}N data.

Introduction

The objective of this paper is to describe in a superficial manner some loss mechanism in lowland rice soils and to document some important problems in the N economy of lowland rice fields. Recent publications [2, 3, 4, 15, 28] have documented the state of knowledge of nitrogen balance studies and hence this aspect will not be reviewed here. The emphasis in this review will be on: (a) loss mechanisms and their dependence on transport processes; (b) yield increases per unit of applied fertilizer N; (c) yields in the absence of applied fertilizer N; and (d) behavior of ^{15}N in the lowland rice soils. Since other papers in this publication will deal with specific mechanisms, the emphasis here will be on observations and not explanations; the questions will be asked here, but such answers as are available will be provided in the succeeding papers.

1

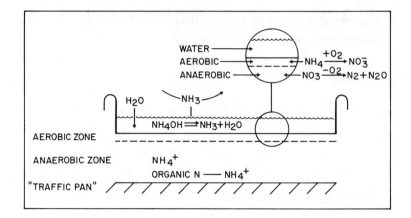

Figure 1. Nitrogen transformations in the ideal paddy.

The ideal lowland rice field

A cross section of an idealized lowland rice field is illustrated in Figure 1 together with the classical N reactions. First, a description of the water regime. In the bunded field shown here and, with ideal water control, there is only minimal and deliberate water movement from the lowland rice field; in some cases water is supplied to a lower field by flow through the upper field but under ideal water control this is managed so that unplanned movement of N is minimal. In an ideal lowland rice field there is a 'pan' at a depth of a few cm which restricts downward percolation (ideally 1 mm/day or less) [18] and lateral movement through the bunds is also not very important. Thus transport of nitrogen by water movement in this ideal lowland rice field is relatively unimportant and the lower limit of the rooting zone is fairly well defined by the 'pan'.

In the ideal lowland rice field the soil above the pan is well puddled and uniformly mixed, resulting in a substrate whose physical strength does not restrict penetration by roots. It is too viscous to be mixed by thermal gradients and, as pointed out above, flow of water is restricted; transport processes are thus mostly restricted to diffusion and mass flow of water.

Illustrated in Figure 1 is the thin oxidized zone (usually 1 to 20 mm in thickness) normally found at the interface between water and soil (5 pp. 89–91, 7, 20, 26, 27). However the bulk of the root zone is devoid of oxygen and in this zone only anaerobic microbial activity is possible. The multitude of chemical changes associated with anaerobic metabolism are all evident in this zone (22–25).

Mineralization of organic nitrogen in the root zone is often appreciable, as evidenced by yields of crops without fertilizer nitrogen and from mineralization in the laboratory (2, 5 p 103, 6, 27). In a later section yields in the absence of fertilizer N will be discussed more completely.

For our purposes we need to emphasize the rapid and complete reduction of NO_3^- to other forms (predominantly N_2 and N_2O) in the anaerobic zone. This precludes nitrate sources of nitrogen in any except extremely unlikely situations. A second important point is that NH_4^+ cannot be oxidized to NO_3^- in the absence of oxygen. The absence of water flow and stability of NH_4^+ means that NH_4^+ sources of N in the anaerobic zone are not subject to serious losses. This applies both to ammonium sources of fertilizer N and to nitrogen mineralized from organic matter.

The losses of N from the ideal lowland rice field as depicted in Figure 1 are largely restricted to the zone adjacent to the interface between water and soil (5 pp 100–103, 20, 21, 27). Here ammonium sources of N may be oxidized to NO_3^- in the thin aerobic soil zone and perhaps in the overlying water. This NO_3^- is then lost from the system when it diffuses to the anaerobic zone where it is denitrified. NH_3 volatilization from the overlying water is also an avenue of loss [17, 27].

So far the major processes mentioned have been the well known ones of mineralization, nitrification, denitrification, leaching and ammonia volatilization. These are the same processes we discuss in upland soils. How then do lowland rice soils differ from upland soils? Basically the biological processes are the same; or at least similar in many respects. Such differences as there are the result of the interactions (or non-interactions) between transport processes and a number of biological and chemical reactions.

The need to consider several processes simultaneously and the importance of transport processes can be illustrated by the ideal lowland rice field described above. First, consider ammonium nitrogen placed at a depth of 10 cm in the lowland rice field which is maintained in a flooded condition. The rate of transport of oxygen into the soil at the water-soil interface is too slow to satisfy the needs of the aerobic organisms and hence only a few mm of the soil contains any oxygen. Since only a mm or so of water is percolating through the paddy and diffusion of NH_4^+ is relatively slow, the bulk of the ammonia nitrogen remains close to where it was placed originally. The lack of transport mechanism for oxygen and the ammonium have prevented most losses of ammonium.

Now let us consider the same nitrogen broadcast on the surface of the bunded lowland rice with no water flow across/through the bunds. As a means of simplifying the argument let us suppose that added urea fertilizer dissolves immediately in the floodwater and hydrolyses immediately to $(NH_4)_2CO_3$. The concentration of NH_4^+-N in solution is inversely proportional to the thickness of the water and directly proportional to the amount of nitrogen added. We must now consider the rate of transfer across the two interfaces (air-water and water-soil) plus whatever reactions occur within the water. At the air-water interface, NH_3 volatilization is the major loss mechanism [17]. Loss is a function of partial pressure of NH_3 in equilibrium with the water, transport in the air-water interface, and transport mechanisms in the air. The partial pressure of NH_3 is in turn set by a number of variables; probably photosynthesis is the most important since during the day it may deplete the CO_2 in the water more rapidly than it can be replenished by transfer across the interfaces. Loss of CO_2 in turn leads to marked increase in pH. Thus we have the sequence:

Photosynthesis depletes CO_2 → increases pH → increases partial pressure of NH_3 → increases NH_3 volatilization.

At the soil-water interface, transfer across the interface will be determined by diffusion and mass flow of water. The major parameters are: concentration gradient, mass flow of water, reactions of ammonical nitrogen with the soil and diffusion coefficient of ammonical species in the soil (5 pp. 105, 20, 21). One interesting aspect of diffusion is that initially the fluxes may be into the soil but as the water is depleted of ammonical nitrogen by volatilization, the flux of ammonia may be reversed from ammonical nitrogen into the soil to ammonical nitrogen back to the water.

Turning attention now to nitrification, depending upon the population of nitrifiers and their build-up, more or less of the ammonical nitrogen will be nitrified in the aerobic zone at the soil water interface. Some of the nitrate will then be transported by diffusion into the overlying water and some will be transported by mass flow and diffusion into the underlying anaerobic soil where it is denitrified (5 pp. 105, 20, 21). Thus we have:

diffusion (mass flow) of NH_4^+-N into soil → nitrification →
diffusion of NO_3^- to anaerobic zone → denitrification

Based on the foregoing idealized lowland rice system, the placement of ammonium sources of nitrogen in the anaerobic zone of the ideal lowland rice field would be expected to be far superior to surface placements and indeed this has been shown to be true where conditions approach the ideal [23].

Based on reasoning about transport processes, soil chemical and microbial processes, and chemical and photosynthetic processes in the water, wide variation among locations and years would be expected from ammonium nitrogen broadcast on the surface of the ideal lowland rice field. Yield response to fertilizer should be inversely proportional to losses and this variation in losses should lead to variaion in yield response. This in fact is observed as will be documented in subsequent discussion.

A common method of nitrogen placement is the so-called 'basal, broadcast and incorporated'. In this procedure ammonium sources of nitrogen are broadcast on the surface of a fairly well puddled field and then incorporated by some tillage operation. A bit of reasoning, imagination and observation of what farmers do, soon leads to the conclusion that variation of losses among farmers, soils, locations, and local conditions must be enormous because of the interactions among the processes which occur (e.g. variation in degree of incorporation might vary from superficial to uniform mixing with several cm of soil).

The non-ideal lowland rice field

Soil properties and water regimes, which differ from the ideal, add additional variation to behavior of soil and fertilizer nitrogen in lowland rice soils. The range of variation in soil and water regimes has been documented and discussed in detail elsewhere [18]. Examples of variations which are likely to have a major impact on behavior of soil and fertilizer nitrogen follow. In light textured soils, the permeability to water may be high enough that some leaching of nitrogen occurs, although this is not well documented. In other situations with high soil permeability, the oxidized zone may be relatively thick (several cm); nitrification may be extensive in the oxidized zone and the flow of water through the soil insures rapid transport of nitrate to the anaerobic zone where it is denitrified. In still other situations the water management may be imperfect and, during the growing season the soil may undergo enough drying that the ammonium nitrogen is oxidized to NO_3^-; upon subsequent reflooding the NO_3^- is denitrified and lost from the system or else leached from the rooting zone. This mechanism (alternating wetting and drying) may be particularly serious sink for soil nitrogen in regions with wet-dry seasons. Following the dry season the soil will often be wet to field capacity by the initial rains at the beginning of the wet season; mineralization and nitrification may occur relatively rapidly under the favorable water and temperature regimes. Then as the rains increase in intensity and frequency the soil becomes water-logged and the previously mineralized nitrogen is denitrified.

Soil properties and water regimes which vary widely from the ideal are very common both on experiment stations and farmers' fields. Only seldom will one find the combination of soil properties and water regimes which approach the ideal. However, the ideal does provide one extremely important piece of information; it demonstrates beyond a shadow of doubt that there is nothing inherent in the lowland rice field regime which precludes high yields per unit of available nitrogen (whether the nitrogen be from soil organic matter or fertilizer). Thus the ideal lowland rice field with proper water control is in fact a situation where extremely high yield increases can be obtained with ammonium sources of nitrogen properly placed (on the order of 50 to 60 kg grain per kg of fertilizer N).

The importance of transport processes

The aim of the above discussion has been to emphasize the linkages between biological and chemical processes and transport processes which determine

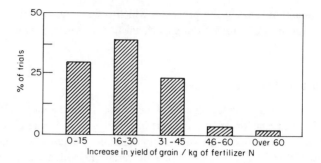

Figure 2. Frequency distribution of grain yield increases per kg of fertilizer N for 109 trials in Asia and Southeast Asia. The increases are for the first increment (28 kg N in wet season, 56 kg N in dry season). Nitrogen was in the form of urea and method of application was considered to be 'best split' by the individual investigators.

the relative importance of the various classical and well known N mechanisms. Furthermore, the differences between upland and lowland soil are not bio-logical processes which are unique to each system but rather the differences are the result of variation in the linkages between the biological, chemical and transport processes which determines the relative importance of the various reactions. In fact one could argue that the kinetics and the relative importance of the various loss mechanisms are mostly determined by transport processes rather than the intensity and capacity of common soil chemical character-istics.

Yield increase per unit of applied fertilizer N

In the final analysis, yield increase of grain per unit of applied fertilizer N is the most important aspect of nitrogen fertilization of rice. Summarized in Figure 2 is a frequency distribution of yield increases reported for the First, Second and Third International Trial on Nitrogen Fertilizer Efficiency in Rice [9–12]. This probably represents the state of the art with conventional sources and methods of application. The source of N was urea and the method of application was what each investigator considered to be the 'best split'. This probably represents something better than farmer management of land preparation, water management, control of pests, variety selection, etc. and we suppose that most farmers would not do this as well as in these experiments.

As an aide in interpretation, 50 kg grain per kg of fertilizer N is considered a practical upper limit [13, 30]. The grain itself will contain about 1.3% N, the straw will contain $\frac{1}{3}$ to $\frac{1}{2}$ as much N as the grain and roots will contain 20 to 30% as much as in the top of the plant. Approximately, the N content of the plant is increased about 1 kg for each kg of added fertilizer N when a

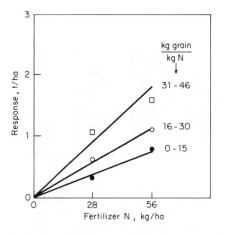

Figure 3. Average response at locations grouped according to kg grain/kg N with 8 locations per groups (International Rice Research Institute 1978b). Points are average of experimental observations, lines are drawn by eye for reference purposes only.

yield increase of 50 kg grain/kg N is obtained. A comparison of this with the results shown in Figure 2 illustrates that only in a very few experiments did the yield increases approach the 'ideal'. In about $\frac{1}{4}$ of the experiments the yield increases ranged from 30 to 45 kg grain/kg N. Thus, there seems no doubt that actual responses were $\frac{1}{2}$ or less of the ideal in approximately $\frac{1}{2}$ of the experiments.

Several possible explanations are:
– losses of fertilizer N by leaching and denitrification of N_2, N_2O and volatilization of NH_3
– immobilization of fertilizer N
– poor root distribution/activity in the zone of soil influenced by fertilizer
– poor response because of low yield potentials

The last possibility is examined by looking at the average response obtained in the wet season of the Second International Trial [10] in which about $\frac{1}{3}$ of each of 24 experiments fell into three response categories. The results shown in Figure 3 illustrate that the average response to two increments of fertilizer was approximately linear and hence there seems no reason to suppose that a major share of the poor response can be attributed to low yield potentials.

Illustrated in Figure 4 is the frequency distribution of responses with different methods and sources of fertilizer N from the three sets of trials referred to for Figure 2. This illustrates the generally improved response obtained with SCU and urea supergranules (USG) relative to the 'best split' and that at almost $\frac{1}{3}$ of the locations, the response to USG fell into the 31–45 kg grain/kg N category. Although the SCU and USG represent an

8

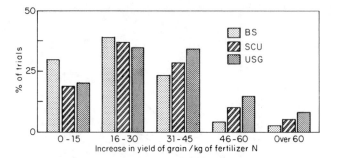

Figure 4. Frequency distribution of grain yield response per kg of fertilizer N for different sources and methods of fertilizer application. B.S. = best split, urea, SCU = sulfur-coated urea, USG = urea supergranules deep placed, 109 experimental observations with B.S., 106 observations with SCU and 74 observations with USG. First increment of fertilizer, 28 kg N/ha in wet season and 56 kg N/ha in dry season.

advancement over the 'best split', at about $\frac{1}{2}$ of the locations the response was on the order of $\frac{1}{2}$ or less of the 'ideal' of 50 kg grain per kg N.

In Table 1, the correlation among the three sources/methods and the means for each is listed. The results indicate that there is considerable difference in behavior among them at the different locations and that response to 'best split' urea and SCU are not correlated with each other even at the 5% level. This indicates two things: (a) that the poor response with 'best split' illustrated in Figure 2 is not entirely a consequence of low yield potential; and (b) whatever cause they behave differently among sources in the same soil. Thus there appears to be no one cause associated with poor response per unit of N for a given soil/experiment, and there are interactions among soils and sources/methods.

There is no evidence from the experiments themselves about the relative importance of three of the factors listed earlier in connection with discussion of Figure 2 (loss of N, N immobilization, ineffective root system). These can only be clarified by studies of soil chemistry, root behavior and crop attributes measured periodically during the growing season. Since there is considerable variability in which sources/method is most effective in a given soil, an association between soil/crop/climate properties and best method/source is essential information in improving response of rice per unit of fertilizer N.

Soil nitrogen

The nitrogen supplied by the soil is sufficient for sizable yields in many situations and in fact, may be equivalent to the yield increases obtained with substantial amounts of fertilizer N (e.g. yields of 3 t/ha without fertilizer N are equivalent to yield increases obtained with 100 kg/ha of 'best split' urea N

Table 1. Correlation matrix and mean responses for 21 experiments in the second international trials (IRRI, 1978b).

	BS	SCU	BQ	Mean kg grain/kg N
BS	1.00	0.37^{NS}	0.53**	29
SCU	0.37^{NS}	1.00	0.49*	33
BQ	0.53**	0.49*	1.00	39

NS	=	Not significant at 5% level
*		Significant to 5% level
**		Significant at 1% level
BS	=	Best split
SCU	=	Sulfur-coated urea
BQ	=	Briquets

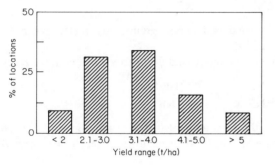

Figure 5. Frequency distribution of yields without fertilizer N in 152 trails in South and Southeast Asia.

when response is 29 kg grain/kg N, the average shown in Table 1). A summary of results from 23 long-term studies showed that 22% of the yields with no fertilizer N ranged between 1 and 2 t/ha, 52% between 2 and 3 t/ha and 26% between 3 and 4 t/ha [15]. Illustrated in Figure 5 are the results from four projects in Asia and SE Asia [9–12] with a total of 152 location-years of results. These results illustrate that approximately 60% of the yields range between 2 and 4 t/ha of rice without fertilizer nitrogen.

The conclusions of this section are the following: (1) the nitrogen supplied by the soil is an extremely important component of rice production; (b) very little research effort is expended on studying how to use soil nitrogen more effectively (or at least reports of extensive research are not found in the literature); and (c) is some fraction of the resources now devoted to fertilizer N were diverted to a study of soil N and its management, then perhaps substantial increases in yield could be obtained with resources already on the land.

Behavior of ^{15}N in lowland rice

The second section cited the lack of information on the general mechanisms responsible for low response per unit of applied N. ^{15}N seems a likely tool to use in such studies and hence the following section is an examination of some uses of ^{15}N reported in the literature. In many ^{15}N studies the response of grain yield per unit of applied N was not a major objective. Thus in this section that parameter will not be used. We will use the parameter 'net change in N in above-ground dry matter per unit of applied N' ($\Delta NP/\Delta NF$) as one parameter and 'change in ^{15}N in above ground dry matter per unit of applied ^{15}N ($^{15}\Delta NP/^{15}\Delta NF$)' as a second parameter.

By definition:

$$\frac{\Delta NP}{\Delta NF} = \frac{NP]F - NP]o}{NF}$$

where $NP]F$ = total N in above-ground dry matter when quantity NF of fertilizer N is added
$NP]o$ = total N in above-ground dry matter when no fertilizer is added

$$\frac{^{15}NP}{^{15}NF} \times 100 = \frac{^{15}NP/f}{^{15}NF}$$

where ^{15}NP is the total amount of ^{15}N in above-ground dry matter, f atom % of ^{15}N fertilizer N and ^{15}NF is quantity of fertilizer tagged with ^{15}N which was added.

In effect, the first quantity is the net effect of the fertilizer addition on the accumulation of N by the above-ground dry matter while the second quantity is the fraction tagged nitrogen which is accumulated by the plant.

Figure 6 summarizes the relation between these two parameters from data found in the literature [1, 3, 8, 14, 16, 19, 29]. Very clearly the uptake of ^{15}N by the plant underestimates by a considerable amount the net effect of the fertilizer on accumulation of N by the plant. The discrepancy between the two is made up of ^{14}N from the soil. The frequency distribution of the ratio of soil N with fertilizer [= total N in plant minus (total ^{15}N in plant/f)] to soil N without fertilizer (= total N in plant when no fertilizer is added) is shown in Figure 7. Thus the addition of fertilizer N markedly enhances the apparent uptake of soil N; in about $\frac{1}{2}$ of the experiments the enhancement was 40% or more.

One interpretation of the latter is that additions of fertilizer will lead to rapid depletion of the soil N (in case the fertilizer N not taken up by the plant is lost from the soil and hence does not balance the enhanced soil N uptake). This interpretation does not seem to be consistent with experience.

Probably at least part of the discrepancy shown in Figure 7 can be attributed to accumulation of more or less ^{15}N by the biomass with release of

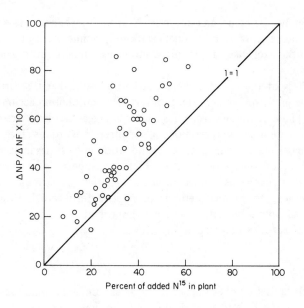

Figure 6. Net increase in total N in above ground dry matter per unit of applied fertilizer N plotted against percent of the ^{15}N added in the fertilizer which is found in the top of the plant. Points are experimental observations, line is 1:1 line drawn for reference purposes only.

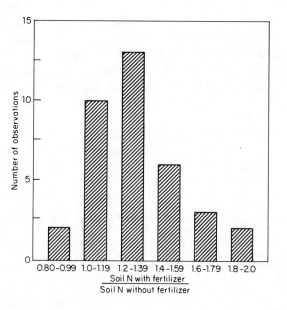

Figure 7. Frequency distribution of ratio of soil N in plant with fertilizer to that in plant when fertilizer is added.

12

approximately the same amount of ^{14}N from the biomass essentially an exchange process, with net mineralization − immobilization being small relative to the exchange of ^{15}N in the inorganis soil pool for (organic) ^{14}N in the biomass.

Regardless of the reasons, there can be no doubt that ^{15}N fertilizers are *not* reliable predictors of *net* effects of fertilizer additions on accumulation of N by the plant. It follows from this that ^{15}N tagged fertilizers are not useful in making economic interpretations of fertilizer N additions since economic yield increases are to some extent dependent upon increases in accumulation of N in the above-ground dry matter. Thus ^{15}N appears to be a useful to the extent that it enables us to differentiate among the various reasons for differential response to sources/methods of applied fertilizer N. However, the results cited above indicate that some important questions remain to be answered about how to interpret ^{15}N data.

Conclusions

1. In a substantial number of location-year experiments, grain yield increase per unit of applied fertilizer N is unacceptably low regardless of source and method of application.

2. The reasons for poor performance appear to vary among sources/methods of application at a given location and among locations.

3. The reasons for poor performance of fertilizer nitrogen can only be determined by study of soil-crop factors during the season at several locations where soil-crop-management factors vary widely among locations.

4. There is an incredible potential for increasing grain yield of rice with no further increases in inputs of fertilizer N if we could learn how to match the best method/source of fertilizer with the soil-crop-management complex.

5. The nitrogen supplied by the soil is an extremely important factor in rice production. More research needs to be devoted to the 'care and management' of soil nitrogen.

6. The interpretation of ^{15}N data seems unclear. However, ^{15}N must be developed as a useful research tool in lowland rice systems in order to achieve the objectives listed above.

References

1. Broadbent FE and Mikkelsen PS (1968) Influence of placement on uptake of tagged nitrogen by rice. Agron J 60:674–677
2. Broadbent FE (1979) Mineralization of organic nitrogen in paddy soils. *In* Nitrogen and Rice. International Rice Research Institute. Manila, Philippines, pp 105–118
3. Cao ZH, De Datta SK and Fillery IRP (1984) Nitrogen-15 balance and residual effects of urea-N in wetland rice fields as affected by deep placement techniques. Soil Sci Soc Am J 48:203–208
4. Craswell ET and Vlek PLG (1979) Fate of fertilizer nitrogen applied to wetland rice. *In* Nitrogen and Rice. International Rice Research Institute. Los Baños, Phillippines

5. De Datta SK (1981) Principles and Practices of Rice Production. John Wiley and Sons, NY pp 89–91
6. De Datta SK, Stangel PJ and Craswell ET (1981) Evaluation of nitrogen fertility and increasing fertilizer efficiency in wetland rice. *In* Proceedings of Symposium on Paddy Soil. Springer-Verlag, New York
7. Howeler RH and Bouldin DR (1971) The diffusion and consumption of oxygen in submerged soil. Soil Sci Soc Amer Proc 35:202–208
8. IAEA (1978) Isotope studies on rice fertilization. International Atomic Energy Agency. Vienna Tech Rep Ser No. 181
9. International Rice Research Institute (1978a) Final report on the first international trial on nitrogen fertilizer efficiency in rice (1975–1976). International Network on Fertilizer Efficiency in Rice (INFER). Los Baños, Laguna, Philippines
10. International Rice Research Institute (1978b) Preliminary report on the second international trial on nitrogen fertilizer efficiency in rice (1977). International Network on Fertilizer Efficiency in Rice (INFER). Los Baños, Laguna, Philippines
11. International Rice Research Institute (1980a) Preliminary report on the third international trial on nitrogen fertility efficiency in rice (1978–1979) International Network on Soil Fertility and Fertilizer Evaluation for Rice (INSFFER). Los Baños, Laguna, Philippines
12. International Rice Research Institute (1980b) Report on the international long-term fertility trial in rice. International Network on Soil Fertility and Fertilizer Evaluation for Rice (INSFFER). Los Baños, Laguna, Philippines
13. Keulen H Van (1977) Nitrogen requirements of rice with special reference to Java. Cont. Centr. Res. Inst. Boger, Indonesia No. 30
14. Koyama T (1971) Soil-plant nutrition studies on tropical rice. III. The effect of soil fertility status of nitrogen and its liberation upon the nitrogen utilization by rice plants on Bangkhen paddy soil. Soil Sci and Plant Nutr 17:210–220
15. Koyama T and App A (1979) Nitrogen balance in flooded rice soils. *In* Nitrogen and Rice. International Rice Research Institute. Los Baños, Philippines
16. Koyama T and Niamscrichand N (1973) Soil-plant nutrition studies on tropical rice. VI. The effect of different levels of nitrogenous fertilizer application on plant growth grain and nitrogen utilization by rice plants. Soil Sci. Plant Nutr. 19:265–274
17 Mikkelsen DS and De Datta SK (1979) Ammonium volatilization from wetland rice soils. *In* Nitrogen and Rice. International Rice Research Institute. Manila, Philippines, pp 135–156
18. Moorman FR and Van Breeman N (1978) Rice: Soil, water, and land. International Rice Research Institute. Manila, Philippines, p 40:91–92.
19. Patnaik S and Broadbent FE (1967) Utilization of tracer nitrogen by rice in relation to time of application. Agron J 59:287
20. Patrick WH and Delaune RD (1972) Characterization of the oxidized and reduced zones in flooded soil. Soil Sci Soc Amer Proc 36:573–576
21. Patrick WH and Reddy KR (1976) Nitrification-denitrification reactions in flooded soils and water bottoms: Dependence on oxygen supply and ammonium diffusion. J Environ Qual 5:469–472
22. Patrick WH and Reddy CN (1978) Chemical changes in rice soils. *In* Soils and Rice. International Rice Research Institute. Manila, Philippines, pp 361–398
23. Prasad R and De Datta SK (1979) Increasing fertilizer nitrogen efficiency in wetland rice. *In* Nitrogen and Rice. International Rice Research Institute, pp 465–484
24. Ponnamperuma FN (1978) Electrochemical changes in submerged soils. *In* Soils and Rice. International Rice Research Institute. Manila, Philippines, pp 421–441
25. Ponnamperuma FN (1982) Some aspects of the physical chemistry of paddy soils. *In* Proceedings of Symposium on Paddy Soils. Science Press Beijing pp 59–94
26. Phuc N, Tanabe K, and Kuroda M (1976) Mathematical analysis on the miscible displacement and diffusion of dissolved oxygen in the submerged soils. J Fac Agric Kyushu Univ 20:61–73
27. Savant NK and De Datta SK (1982) Nitrogen transformation in wetland rice soils. Adv Agron 35:241–302

14

28. Watanabe I, Craswell ET and App AA (1981) Nitrogen cycling in wetland rice soils in East and Southeast Asia. *In* Wetselaer R et al. (eds) Nitrogen Cycling in Southeast Asian ecosystems. Australian Academy of Science. Canberra
29. Yoshida T and Padre BC (1977) Transformation of soil and fertilizer nitrogen in paddy soil and their availability to rice plants. Plant and Soil 47:113–123
30. Yoshida S, Forno DA, Cock JH and Gomez KA (1976) Laboratory manual for physiological studies of rice. 3rd ed. International Rice Research Institute. Los Baños, Philippines

2. Nitrogen transformations in flooded rice soils

DR KEENEY and KL SAHRAWAT

Department of Soil Science, University of Wisconsin-Madison, Madison, WI 53706, USA

Key words: lowland rice, mineralization-immobilization, denitrification, biological
reactions, dissimilatory nitrate reduction

Abstract. A review is made of the recent literature pertaining to the reactions and
processes that soil and fertilizer N undergo in lowland rice soils in relation to the
improved N management and overal N economy of lowland rice soils. Topics discussed
include: nitrogen leaching, ammonium fixation and release, ammonia volatilization,
N_2 fixation, mineralization-immobilization, nitrification-denitrification, dissimilatory
nitrate reduction, urea hydrolysis, critical pathways for control of nitrogen loss.

Flooded soils differ considerably from their arable counterparts in several
characteristics. Perhaps the characteristic that makes the flooded soils
markedly different from arable soils, and which also greatly affects N transformations and fertilizer use by rice is their low supply of O_2. Thus, they are
reduced most of the season, the anaerobic metabolism is dominated by
bacteria, and the products of metabolism differ markedly from the arable
soils [42, 67, 93, 112, 113].

The presence of oxidized and reduced soil layers (see Figure 1) makes the
flooded soils a unique system where both aerobic and anaerobic N metabolism
can occur in close proximity. Thus N is markedly susceptible to losses in
these soils (Table 1).

Several reviews are available that discuss various aspects of the N cycle in
flooded soils and sediments [7, 8, 22, 42, 59, 60, 67, 71, 75, 92, 93, 105,
112, 113]. We will focus on N transformations and transport processes in
flooded soils that have relevance to improved N management and overall N
economy of lowland rice soils. The interest in N transformations in flooded
soil ecosystems stems from the fact that rice, which is the staple food for half
of the world population [15, 22], does not use fertilizer N very efficiently
[16, 69, 75, 89].

Nitrogen is the nutrient element limiting growth in most rice-growing soils
[92]. Further, increased yields due to improved management involves use of
fertilizer N. Better understanding of the availability of N from the soil organic
N and the fate of added N fertilizer should aid in development of innovative
N management technology. Even a small increase in the efficiency of fertilizer

15

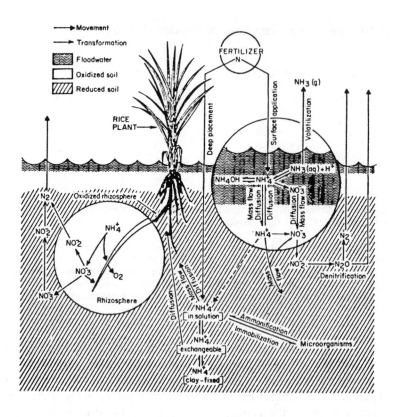

Table 1. Biochemical nitrogen transformation reactions that occur in the different redox zones of an idealized flooded soil-water system

Zone	Redox state	Dominant nitrogen transformation reactions
Floodwater	Oxidized	N_2 fixation by algae, aerobic bacteria; nitrification, ammonia volatilization
Oxidized surface face layer	Partially oxidized	Ammonification, nitrification, immobilization, N_2 fixation by algae, bacteria
Reduced soil	Reduced	N_2 fixation, ammonification, immobilization, denitrification, (reductive deamination), dissimilatory nitrate reduction
Rhizosphere	Partially oxidized	N_2 fixation, ammonification, nitrification, (oxidative deamination), defitrification

N will save energy costs and foreign exchange spent by countries where N fertilizer must be imported.

Physical/metabolic zones of flooded soils

A flooded soil is a dynamic heterogeneous soil-water system that has three distinct soil layers established mainly by the prevailing oxidation-reduction or redox potential (Eh or pE) of the system. The floodwater and a few mm to one cm thickness of the surface soil in contact with the water is partially aerated, usually has a relatively high redox potential and supports aerobic microbial reactions (Table 1). The pH of the overlying water phase rapidly fluctuates diurnally in response to algae growth. Removal of CO_2 during photosynthesis results in marked increase in pH of the floodwater, and volatilization of NH_3, if NH_4^+ is present, can result in significant N losses [51, 52, 108]. The plow layer of a flooded lowland usually is several cm thick and has a low Eh or pE (pE = $-$ log e = Eh/0.059) conductive for NH_4^+ accumulation. The presence of the oxidized zone in close proximity to the reduced soil zone is also conductive for the loss of N through nitrification (oxidized layer) followed by denitrification (reduced zone) [72, 73].

Ponnamperuma [67] described the NO_3^-–N_2 system by the following equation, which indicates that in the flooded soil, where pE may range from -1 to 3, NO_3^- is extremely unstable:

$$pE = 21.06 - 1/5 \ pNO_3^- + 1/10 \ pN_2 - 6/5 \ pH$$

The instability of NO_3^- in flooded soils has been long recognized [64] and its loss via denitrification when applied to wetland rice soils has also indirectly been recognized by the poor performance of NO_3^- fertilizers as a N source for lowland rice [19, 41].

The rhizosphere of a lowland rice plant is partially oxidized due to entry of O_2 to rice roots through the rice aerial parts. Savant and DeDatta [92] reported that the apparent pE of the rhizosphere of 5 to 6 week old IR36 rice plants growing in a reduced clay soil (pE = 0 to -3) ranged from $+2$ to $+5$. Thus, rhizosphere of a submerged lowland rice field may support aerobic N reactions such as nitrification, mineralization of organic N via oxidative deamination and biological N_2 fixation by aerobes and facultative anaerobes.

Physical and chemical processes

Nitrogen movement and distribution

Ammonium can be leached more readily in a reduced than in an arable soil. The rate of movement increases as the pE of the soils declines and is the result of the release of cations such as Fe^{2+} and Mn^{2+} that compete with NH_4^+ on the exchange sites [68].

Nitrate leaching may be prevalent in light-textured (sandy) soils that are hard to maintain in the flooded state. Examples of these soils are found in Punjab (India) where high percolation rates result in large losses of deep point-placed N. A greenhouse study by Vlek et al. [107] provided evidence of high N loss due to leaching in soils with low CEC and high percolation rates. Savant and DeDatta [92] have summarized the recent information on movement and leaching of N in lowland rice soils. The puddling of a soil and its compaction should reduce greatly the rate of water movement and thus N leaching. However, there is very little data on N leaching of flooded soils under field conditions. In a recent field study in Louisiana using ^{15}N fertilizers, negligible amounts of N moved beyond the 20-cm depth in a flooded Crowley silt loam [71]. Similarly, Savant and DeDatta [91] reported that NH_4^+ formed from surface applied urea had moved 12 to 14 cm in a submerged undisturbed clay soil by 4 weeks after application of urea in the absence of rice plants. The movement of NH_4^+ in a lowland field was: downward > lateral > upward from the deep placement (10 cm) of urea. The NH_4^+ concentration gradient disappeared earlier in the dry than in the wet season, probably due to faster movement of NH_4^+ and/or greater root sink effect in the dry season [91, 92].

Ammonium fixation and release

Less emphasis seems to have been given to the dynamics of NH_4^+ fixation and release in flooded soils than in arable soils. This is mainly due to the generally accepted belief that NH_4^+ fixation is not of any significance in lowland rice soils. For example, it is generally stated that the 2:1 type clay minerals that are known to entrap NH_4^+ in arable soils do not fix NH_4^+ in flooded soils because fixation is usually associated with drying to moisture contents usually not relevant to flooded soils. However, it is known [7] that soils containing significant amounts of vermiculite and illite are capable of fixing NH_4^+ under moist conditions [76]. Moreover, recent studies have indicated that NH_4^+ fixation is important even in lowland rice soils [44, 77]. Also, flooded soils are often drained and used for rice-based cropping systems where the second crop is grown under upland conditions. It was further shown in a study with 12 diverse tropical rice soils that these soils fixed NH_4^+ when treated with $(NH_4)_2SO_4$ solution under flooded conditions. The NH_4^+-fixing capacity of the soils ranged from 3.8 to 7.7 meq/100 g of soil. Ammonium fixation in these soils was not related to pH, organic matter, or clay content but was significantly correlated ($r = 0.61^*$) with the amount of active iron [77]. It was suggested that because of the reversible oxidation and reduction of iron oxides in rice soils, this mechanism of NH_4^+ fixation may be of special importance in sorption and desorption of NH_4^+ and its availability to rice. It was also found that the oxidation of organic matter by hydrogen peroxide in Maahas clay (the major soil series at the IRRI farm) doubled NH_4^+ fixation probably due to exposure of fresh NH_4^+ fixing sites. Similarly, a recent study with the clay fractions separated from 14 lowland Philippine rice soils showed that beidellitic and vermiculite clays fixed more than 90% of the applied NH_4^+ under hydromorphic conditions, while a montmorillonite clay fixed 50% of the applied NH_4^+. Clays of all other mineralogical compositions containing chlorite, hydrous mica, halloysite, kaolinite and amorphous materials did not fix significant amounts of ammonium [5].

Tilo et al. [104] studied the distribution of native fixed NH_4^+ in the profiles of 16 Philippine soils including some used for lowland rice. Fixed NH_4^+-N ranged from 7 to 428 mg/kg of soil and constituted 1 to 56% of the total N. The surface sample collected from a lowland rice field contained the highest concentration of fixed NH_4^+-N (428 mg/kg) and this comprised 17.9% of the total soil N. These and other studies [44] clearly indicated the potential importance of fixed NH_4^+ in the N cycle in flooded rice soils. Better understanding of the dynamics of fixation and release of NH_4^+ is highly desirable for its relevance to N management of lowland rice. Results of a recent greenhouse pot study indicated that the release of fixed NH_4^+ under submerged conditions of rice culture may be faster and more significant than commonly reported for arable soils [48]. Using ^{15}N-labeled ammonium sulfate fertilizer, it was found that the residual fixed NH_4^+ decreased from 45 to 23% during cropping with rice under flooded conditions. The dynamics of NH_4^+ and its fixation in flooded rice soils is further discussed by Mengal et al. (this volume).

Nommik and Vahtras [57] have comprehensively discussed the retention and fixation of NH_4^+ in soils, covering mainly the arable soils. It was suggested that the question of availability of interlayer fixed NH_4^+ in soils cannot be fully resolved by the nitrification test or by chemical laboratory tests used for determining NH_4^+ fixing capacity of soil in relation to its availability to field crops. Fixation of NH_4^+ may be a desirable factor in preventing loss of N, thus ensuring sustained supply of N to plants in a growing season [7]. This hypothesis has been confirmed by field studies by Keerthisinghe et al. [44].

Ammonia volatilization

Loss of nitrogen through NH_3 volatilization from soils including flooded soils has been a subject of several recent comprehensive reviews [51, 55, 102, 108]. Additionally, this issue has been dealt in reviews on N transformations by several authors [7, 22, 34, 52, 59, 75, 92]. Fillery and Vlek (this volume) have reported the significance of NH_3 volatilization as a N loss mechanism in flooded rice soils. We briefly cover the salient principles relevant to NH_3 volatilization from flooded rice fields. It is clearly evident from literature that estimates of the magnitude of NH_3 volatilization loss may vary widely with the technique used for its measurement [22, 92].

Of the several factors that affect NH_3 volatilization, the pH of the floodwater has been recently recognized as the single most important determinant [52]. However, its importance in aquatic systems and its sensitivity to CO_2

concentration as a result of photosynthetic activity has long been recognized [58]. The pH of the floodwater of a flooded soil follows diurnal fluctuations and may increase or decrease by two units during the 24-hour period in response to photosynthetic activity of biota and temperature [51, 52]. Ponnamperuma [68] suggested that the pH of floodwater was related to CO_2 concentration and HCO_3^- activity:

$$pH = 7.85 + \log(HCO_3^-) - PCO_2$$

Thus high bicarbonates in a system with constant removal of CO_2 may greatly increase the pH which can increase NH_3 volatilization of surface applied fertilizer or of NH_4^+ which diffuses into the water layer. The fluctuation in the floodwater pH is further governed by the buffering capacity of the flooded soil-water system.

Floodwater pH is the result of interactions of several floodwater properties including concentration of dissolved CO_2 and NH_3, pH buffering capacity, alkalinity, temperature and biotic activity. Several other factors involving the soil (pH, CEC, PCO_2, buffering capacity, and alkalinity) and the environment (temperature and wind velocity, etc.) as well as the nature and amount of fertilizer N applied and size of plant canopy affect NH_3 volatilization loss from a flooded soil [92].

In general, losses of NH_3 are higher in alkaline and calcareous soils and increase with an increase in soil pH, temperature and solar radiation but decrease with an increase in CEC of the soil and other cultural and management practices including the presence of rice canopy activities which decrease the amount of NH_3 in solution. Also, higher losses of volatile NH_3 are reported from urea fertilizer compared to other NH_4^+ sources because hydrolysis of urea provides alkalinity which can maintain or initiate volatile loss of NH_3.

Volatilization of NH_3 generates protons [4] which tend to acidify the system and will eventually retard loss unless there is constant supply of alkalinity (e.g., by urea hydrolysis). Application of N fertilizer in the reduced layer or to the crop when its root system is well established apparently curtails these losses because both practices decrease the amounts of ammonium that is available for volatilization [18].

Biological processes

Nitrogen fixation

Flooded soils are an ideal habitat for N fixation by nonsymbiotic, anaerobic and aerobic microbes. This can contribute significantly to the N nutrition of lowland rice [12, 20, 67, 110]. Nitrogen fixation is greater in flooded than in upland soils [114, 115]. This topic is covered in detail by Roger and Watanabe (this volume).

Table 2. Range and mean values of ammonification rates in 39 Philippine lowland rice soils at two temperatures as determined by anaerobic incubation tests[a]

Incubation temperature	Period of incubation	Rate of NH_4^+-N production (mg NH_4^+-N kg dry soil^{-1} day^{-1})	
(°C)	(days)	Range	Mean
30	14	1.2–30.6	5.6
40	7	1.9–74.6	14.0

[a]Calculated from Sahrawat [82]; soils had a wide range in pH (4.3 to 7.9), organic C (0.63 to 5.46%) and total N (0.06 to 0.60%) contents.

Mineralization-immobilization

Mineralization and immobilization processes occur simultaneously in flooded soils with their rates and magnitude influenced by soil and environmental factors [7, 59, 92]. Both oxidative and reductive deamination processes contribute to ammonification in flooded soils. Lack of oxygen supply generally inhibits nitrification and greatly influences the rate of ammonification.

Mineralization of organic N to NH_4^+ is the key process in the N nutrition of lowland rice [7, 8, 40, 59, 85, 86, 92]. Important environmental factors that affect mineralization-immobilization are temperature, soil moisture regime, and soil drying; soil characteristics include pH, organic matter content, C/N ratio, and amount and quality of organic residues.

Net mineralization of soil organic N in four Philippine soils under anaerobic incubation increased with an increase in temperature from 15 to 45°C; the Q_{10} for ammonification ranged from 1.0 to 1.8 [36]. Numerous other studies also emphasize the importance of temperature on the rate of net N mineralization in flooded soils [7, 28, 85, 92]. In a recent study of 39 diverse Philippine lowland rice soils, Sahrawat [82] found that the mean rate of NH_4^+ production increased from 5.6 to 14.0 mg NH_4^+-N kg dry soil^{-1} day^{-1} when the incubation temperature was increased from 30 to 40°C (Table 2). These findings indicate that the temperature prevalent during the growing season should be considered when assessing the N supplying capacity of lowland rice soils.

Immobilization is also a temperature-dependent microbial process and under conditions favorable for N immobilization (application of high C/N ratio residues), immobilization also increases with an increase in temperature.

Drying of soils enhances the N mineralization rate [94–96]. For example, a marked effect of soil drying was observed in four permanently waterlogged histosols in the Philippines [81]. Nitrogen availability to wet season rice was affected by the dry season soil conditions [106].

Among the soil characteristics, organic matter content as measured by organic C and total N account for the most variation in NH_4^+ production under anaerobic incubation. In a recent study, Sahrawat [85] reported that NH_4^+ production in Philippine lowland soils under anaerobic incubation was

22

Table 3. Distribution of mineralizable N in 39 lowland rice soils in relation to total N and organic C content[a]

Mineralizable N[b] (mg kg^{-1} dry soil)	No. of samples	Associated soil properties	
		Total N (%)	Organic C (%)
50	24	0.06−0.16	0.63−1.15
50−100	7	0.16−0.21	1.48−2.14
100−200	4	0.16−0.26	1.97−2.50
200	4	0.31−0.60	2.44−5.46

[a]From Sahrawat [85].
[b]NH$_4^+$-N released under anaerobic incubation of soils at 30°C for two weeks.

highly correlated with total N (r = 0.94**), organic C (r = 0.91**) and C/N ratio (− 0.46**), but was not significantly correlated with CEC, clay or pH. Multiple regression analysis of CEC, pH and clay on mineralizable N accounted only for 36% of the variability. While soil properties such as pH, clay and CEC may be related to N mineralization, their individual contribution to this process could not be clearly quantified because of the numerous interactive effects and cross-correlations of these properties. The association of organic C and total N with mineralizable N in 39 soils studied is evident from data in Table 3.

Liming has been reported to increase the availability of N in flooded soils and its availability to lowland rice [2, 6, 65]. The effect of pH per se cannot be evaluated from such studies. However, a recent investigation showed that net N mineralization occurred in the two acid sulfate soils from the Philippines having a pH of 3.4 and 3.7, respectively [78] (Table 4). It would appear from this study and other evidence that ammonification seems to operate under a wide pH range in flooded soils [85], although the tendency of pH to approach neutrality might also be a factor.

In addition to soil and environmental factors, the quantity and quality (C/N ratio) of organic residues added also affect the release of NH$_4^+$ in submerged soils. Earlier researchers realized that the 'N factor' commonly used for characterizing the N immobilizing capacity of the decomposing residues is lower for flooded soils than for the aerobic incubation [3]. Thus it follows that organic residues with similar C/N ratio will immobilize less N, and the net release of N from these will occur at a relatively higher C/N ratio under flooded than under nonflooded, aerobic conditions. This is supported by results from field studies [111].

Ammonification is also affected by tillage and other operations used for preparation of lowland rice fields [28], but it is difficult to quantify the positive effects of these practices because puddling of soil affects N utilization by lowland rice [23] in ways other than by enhancing mineralization (for example, lessening the movement of N) [90].

Mineralization of soil N is also affected by the presence of the rice plant. For example, Broadbent and Tusneem [9], in a greenhouse study using ^{15}N

Table 4. Mineralization of soil organic nitrogen under anaerobic incubation at $30°C$ for two weeks in two acid sulfate soils from the Philippines[a]

Soil	pH (1:1 H_2O)	Organic C (%)	Total N (%)	NH_4^+-formed (mg kg dry soil^{-1})
Calalahan sandy loam	3.4	1.57	0.110	83
Malinao loamy sand	3.7	1.22	0.090	72

[a]From Sahrawat [78].

fertilizer calculated the apparent net mineralization of soil N from soil N uptake in a flooded Maahas clay (Andaqueptic Haplaquolls). They found that soil N mineralization was higher in the presence of the rice plant than in the unplanted soil because the presence of active rice roots decreased N loss due to nitrification-denitrification. They felt the observed pattern of N mineralization was more closely related to the actual field situations than in incubation tests where the NH_4^+-N accumulation peak tends to level off or decrease with time.

Studies on N immobilization by rice straw under flooded conditions indicate that the fertilizer N was mainly immobilized into the α-amino N fraction and a good part of the immobilized N was remineralized under subsequent anaerobic incubation [105].

Nitrogen release in relation to plant needs

Mineralization of soil organic N in flooded soils is the key process for N nutrition of lowland rice. Even in well-fertilized lowland rice fields, rice utilizes 50–75% of soil N through mineralization [7, 35, 46, 86].

Studies indicate that much of the mineralizable N in a flooded soil is released as NH_4^+ within two weeks of flooding provided temperature is favorable and the soil is neither strongly acid nor greatly deficient in available P [67]. The release of NH_4^+ in laboratory incubated flooded soils follows approximately an asymptotic curve [66]. This NH_4^+ release pattern may not be ideally suited to the N needs of lowland rice because N uptake by rice follows a sigmoidal curve [37].

As pointed out by Broadbent [7], incubation tests may at times give misleading N release patterns because during these test NH_4^+ production, after reaching a peak, tends to level off as early as 2 to 4 weeks of incubation. Nitrogen uptake data under field conditions using ^{15}N fertilizer, however, indicate constant supply of soil N throughout the growing season. If incubation tests are to be useful in predicting the N supplying capacity of lowland rice soils, the pattern of NH_4^+ release should be, in theory, similar to the N release pattern in the field in the presence of rice plants. It is possible that if the NH_4^+ released during anaerobic incubation of soil were periodically removed to simulate N uptake by the rice plant a better characterization of the N supplying capacity of lowland rice soils would result. This is technically very difficult. Comparison of N release in laboratory incubation and N mineralization under field conditions during a growing season should give useful leads in devising and standardizing incubation tests for realistic estimate of the N supplying capacity of a soil. Such studies should also provide information regarding factors that should be considered for modeling of the N cycle. No such studies have been attempted for flooded rice soil but reports comparable to those used for arable soils have been published [39, 101].

Prediction methods

The inefficient use of fertilizer N and heavy dependence by rice on the soil mineralizable N pool emphasizes the need for methods to assess the N supplying capacity of lowland rice soils. Recently, Sahrawat [86] has reviewed the available information about the methods currently used for predicting N availability to lowland rice. Among the biological indices used, anaerobic incubation methods involving incubation of soils under waterlogged conditions at 30°C for two weeks or at 40°C for 1 week are regarded as most useful in predicting the soil N availability to lowland rice. Most of these evaluations have involved greenhouse trials, but there were also a few field tests. These indices would likely be more useful if the temperature prevalent in the region during the growing season were used.

Among the chemical indices, organic C content has been widely evaluated for predicting N availability to submerged rice especially in India [see ref. 86 for review]. This method has been more successful in predicting N availability to rice in greenhouse than in the field situations. However, recent work suggests that this simple test could be made more useful if some component pertaining to the quality of organic matter is also incorporated. The characterization of quality of organic matter might help in explaining the difference in the amounts of N released in soils with the same content of organic matter. Chemical characterization of the soil organic N pool in some Philippine lowland rice soils using alkaline permanganate, acid permanganate, acid dichromate, H_2O_2 and acid hydrolysis suggests that it may be possible to quantify the fraction of soil organic matter which is the source of mineralizable N [82]. This work led to the development of a simple method based on modification of the Walkley–Black (acid dichromate oxidation) method of organic C determination, which can be used for simultaneous determination of organic C and potentially mineralizable N in soils [83]. This method offers an opportunity to test a combination of total organic matter and mineralizable N for predicting N availability to lowland rice.

Among the chemical methods, the one based on the measurement of NH_4^+ released during the digestion of soil samples with alkaline permanganate for a brief period has been widely tested in India for predicting soil N availability. Results, however, have been mixed [86]. Recent research on this method has improved our knowledge about its chemistry [88, 89]. A study by Sahrawat and Burford [88] suggests that this method is a relatively poor predictor of N availability to crops grown in arable soils because of its inability to include NO_3^--N in the available N pool. It is much better for submerged rice, where NH_4^+ is the dominant mineral N form and NO_3^- contributed little to N nutrition.

Greenhouse studies with submerged rice using diverse soils suggest that the chemical methods based on the release of NH_4^+-N from soils by the oxidative action of acid permanganate, acid dichromate and hydrogen peroxide are relatively good predictors of N availability [86].

Recent studies employed the electroultrafiltration (EUF) technique [56] for fractionation of soil N into N fractions which are in soil solution (intensity) or in soil reserve (capacity) by using varying voltage and temperature. This research suggests that EUF-NH_4^+, which comes in soil solution (fraction I) at low voltage (intensity factor), is a good measure of readily available N to lowland rice [50].

The A-value concept [33] has been evaluated in several field studies. Different workers have found that the A-value of a soil varies not only with interactions of fertilizer N with rice but also with the method, rate, and time of fertilizer N application [7, 34, 46]. However, under well-characterized conditions, this method could be of utility in assessing the N supplying capacity of lowland rice soils. With the availability of ^{15}N depleted N fertilizers, this method may prove less expensive and in need of further evaluation. Results obtained with A-values for lowland rice are summarized by Sahrawat [86].

Importance of temporary immobilization

Immobilization is a key process in the N turnover in lowland rice soils, especially in situations where organic residues or manures are used as N sources. Organic N and mineral N pools in a soil are in dynamic equilibrium and the net effects of factors which affect mineralization-immobilization reaction govern the availability of N to plants. As Kai and Wada [40] state, our knowledge regarding the immobilization process in lowland rice is limited compared to what is known in arable soils. They posed three questions: (1) What is the mineralization pattern of native soil organic N and of the recently immobilized N? (2) How long is the immobilized N tied up before it is remineralized? (3) How effectively and efficiently are soil organic N and immobilized N recovered by the rice crop?

These questions cannot be satisfactorily answered because the behavior of immobilized N in lowland rice culture is not fully understood. However, recent laboratory and greenhouse studies using ^{15}N fertilizer suggest that remineralization of immobilized N is slower under flooded soils [40, 48, 105]. Immobilized N acts as a slowly available N source and at times may be helpful in locking up mineral N from physical and biochemical reactions in soils which lead to N loss. We have to learn more about biological N immobilization to appreciate its effects on N economy in lowland rice soils.

Nitrification

General

Nitrification, a strictly aerobic microbial process, occurs in the oxidized surface layer of a flooded soil. However, it is difficult to study nitrification *in situ* in a flooded soil system because as soon as NO_3^- is formed it diffuses down to the reduced layer and is lost from the system by denitrification or reduced to NH_4^+ by dissimilatory NO_3^- reduction [7, 10, 30, 92, 105]. Thus, it is not surprising that the occurrence of nitrification in the oxidized soil layer has been difficult to document. However, occurrence of nitrification is recognized as a mechanism of N loss via nitrification-denitrification in flooded soils and has led to the conclusion that NO_3^- is an inefficient source of N for submerged rice culture [1, 19, 21, 41, 53, 61].

In a laboratory experiment using ^{15}N labeled $(NH_4)_2SO_4$, Yoshida and Padre [116] found that the oxidized layer of a clay soil had high nitrifying activity. After 30 days, nearly one-third of the NH_4^+-N applied $(400 \, mg \, kg^{-1}$ NH_4^+-N) was converted into NO_3^- $(123 \, mg \, kg^{-1}$ NO_3^--N) at 20°C and was detected in the soil solution. A pure strain of *Nitrosomonas europaea* added to an autoclaved soil resulted in oxidation of NH_4^+ to NO_2^--N. This indicated that nitrifiers are active in submerged soils. Nearly one-fourth of the NH_4^+-N applied was converted into NO_3^--N under flooded conditions at 30°C.

Reddy et al. [74] reported that the net nitrification rate in the oxidized surface layer of a flooded soil was $2.07 \, mg \, NO_3^-$-N kg^{-1} soil day^{-1}. Its occurrence and extent was controlled by oxygen diffusion rates, NH_4^+ concentration, thickness of the oxidized layer, and the levels of inorganic C [71].

Occurrence of nitrification in the rhizosphere of a rice plant, which Savant and DeDatta [91] referred to as site II, growing under flooded conditions is a subject of speculation as much as the oxidized or reduced state of the rhizosphere itself. No data are available on the occurrence of in situ nitrification in the rhizosphere of a flooded rice soil.

Problem soils

Sahrawat [84] studied the nitrification of soil N in several problem rice soils having a wide range of pH (3.4 to 8.6) and organic C (1.22 to 22.70%) by incubating them under aerobic conditions for 4 weeks at 30°C. It was found that the two acid sulfate soils (pH 3.4 and 3.7) and an acid soil (pH 4.4) did not nitrify during this period. Mineral and organic soils having pH > 6.0 nitrified at rapid rates and accumulated NO_3^--N ranging from 98 to 123 mg kg^{-1} of dry soil. Alkalizing a near-neutral clay soil by adding 13 g kg^{-1} Na_2CO_3 increased the soil pH from 6.5 to 8.6 but the amount of NO_3^--N produced increased from only 106 mg to 118 mg NO_3^--N kg^{-1} of soil.

Nitrification in these soils, as measured by NO_3^- accumulation, was highly significantly correlated with the soil pH ($r = 0.86^{**}$, $n = 10$), but was not significantly correlated with their organic C or total N contents. However, no

Table 5. Depletion of carbon, total N and different soil N fractions in nine organic soils during six months of incubation at 30°C under aerobic or waterlogged conditions[a]

| Incubation | C | Total N | Nonhydro-lyzable N | Hydrolyzable N | | | |
				Ammonium	Hexos-amine	Amino acid	Unidenti-fied[b]
				% loss on incubation[c]			
Aerobic	18.7	20.1	− 171.5	13.4	44.6	19.1	49.3
Water-logged	18.2	16.2	− 188.4	9.8	47.8	18.0	40.2

[a]Adapted from Isirimah and Keeney [38]. Results reported are average for nine soil samples.
[b]Unidentified N = total hydrolyzable N − (ammonium + hexosamine + amino acid N).
[c]$\dfrac{\mu g\,N\,g^{-1}\ soil\ in\ original\ sample\ fraction - \mu g\,N\,g^{-1}\ soil\ in\ incubated\ sample}{\mu g\,N\,g^{-1}\ soil\ in\ original\ sample\ fraction} \times 100.$

significant correlation existed between nitrification and soil pH when six soils having pH > 6.0 were considered, indicating that increase in soil pH beyond 6.0 did not significantly affect nitrification. It is known that NH_4^+ oxidation is slow in soils having pH lower than 5.0 but increases with increase in pH up to 8.0, although the rate of NO_2^- oxidation is greatly retarded at high pH because of toxicity of free NH_3 to *Nitrobacter* [30].

Isirimah and Keeney [38] studied N transformations in nine organic soils from Wisconsin by incubation under aerobic or waterlogged conditions at 30°C for six months. Mineralization was faster in the more decomposed histic materials. The rate of decline in total organic C and N of the samples was similar under the two moisture regimes. On the average, 20 and 16% of the N was lost from the soil organic pool under aerobic and anaerobic conditions, respectively. Much of the mineralizable N released in these soils during incubation was derived from the acid-hydrolyzable organic N, largely the hexosamine-N, amino acid-N, and unidentified-N fractions. Microbial turnover of hydrolyzable N to refractory (nonhydrolyzable) N fractions was evident (Table 5). These results suggest that the unidentified soil N fraction and the hexosamine fraction contributed most to the mineralizable N pool under aerobic and anaerobic incubation.

Control of nitrification

Nitrification is at low ebb in soils having pH lower than 5, and an acid soil ecosystem is a deterrent to nitrification and its subsequent loss. But since the pH in reduced flooded soils tends to converge to near neutral, nitrification is likely to occur in acid soils which are kept flooded for prolonged periods and have enough organic matter to effect reduction.

Placement of fertilizer N in the reduced zone of a flooded soil reduces nitrification. While the NH_4^+ formed may diffuse to the oxidized layer, the amount susceptible to nitrification will be much less than if N fertilizer is applied to the surface. Also, application of fertilizer N when the rice root system is established and N is being rapidly taken up greatly reduces the availability of NH_4^+ for nitrification [7, 92].

Use of nitrification inhibitors, such as nitrapyrin or dicyandiamide, should be helpful in retarding nitrification, particularly in lowland rice fields where the moisture regime is fluctuating. Application of a nitrification retarding chemical at the site where nitrification occurs should be the most effective way of controlling nitrification [80]. Recent literature on the use of nitrification inhibitors and slow release N fertilizers for lowland rice soils is summarized by Prasad and DeDatta [69]. Sahrawat [80] and Mulvaney and Bremner [54] have discussed the potential of regulating the nitrification process in soil with the use of chemicals, and most of the recent literature on nitrification inhibitors can be found in these reviews.

Denitrification

General

Flooded soils adequately supplied with organic matter under warm climate provide a conducive environment for denitrification loss if the substrate NO_3^--N is available. Until recently [17], most of the denitrification loss estimates were made indirectly by the N balance approach. Thus the measured loss due to denitrification ranges widely. The denitrification process in soil ecosystems has been the subject of several excellent recent reviews [26, 29, 31, 62]. Focht [30], Patrick [59], and Savant and DeDatta [92] have covered the aspects of denitrification relevant to the mechanism of N loss in lowland rice soils. We will briefly discuss the recent work on the direct measurement of denitrification in flooded soils.

Several factors including soil pH, organic matter content, temperature, O_2 diffusion, and nitrification rate affect the denitrification rate in a flooded soil. Broadbent and Tusneem [9], using ^{15}N labeled $(NH_4)_2SO_4$, demonstrated that $^{15}NH_4^+$-N underwent nitrification and denitrification in flooded soil. They further found that when O_2 was absent in the system, no loss of $^{15}NH_4^+$-N occurred. This study provided direct evidence of the occurrence of concurrent nitrification-denitrification in a flooded soil. The loss of $^{15}NH_4^+$-N by denitrification as N_2 was 9.3% in an O_2 atmosphere but only 0.2% in an anaerobic (100% Kr) environment (Table 6). No labeled N_2O was detected. Growing rice plants markedly lessened the extent of N loss. However, inhibition of nitrification with nitrapyrin did not lessen loss of the surface applied $^{15}NH_4^+$-N.

Denmead et al. [27] reported that the loss of N as N_2O from a flooded field containing $40\,kg\,NO_3^-$-N ha^{-1} in the surface soil (pH 5.8) was only 1.4% of the apparent loss. Similarly, Smith et al. [97] found that the loss of urea N (90 and $180\,kg\,N\,ha^{-1}$) applied to lowland rice as N_2O represented only 0.01 to 0.05% of the urea-N applied. Freney et al. [32] studied the loss of N as N_2O following applications of $(NH_4)_2SO_4$ to flooded rice in the Philippines

Table 6. Distribution of ^{15}N in various fractions in a Sacramento clay after 24 days of incubation under flooded conditions as affected by composition of atmosphere[a]

Composition of incubation atmosphere	NH_4^+-N	NO_3^--N	Organic + clay-fixed N	N_2	Total
		(μg g^{-1} of soil)			
100% O_2	4.9	8.1	68.4	9.3	90.7
30% O_2-70% Kr	13.7	0.4	8.6	1.2	96.9
100% Kr	13.3	0.2	83.9	0.2	97.6

[a]From Broadbent and Tusneem [9].

and reported that N_2O losses were only 0.1% of the 120 kg N applied. Similar low values of N_2O losses were reported by Craswell and DeDatta [17].

These studies suggest that N_2O is not a significatnt gaseous product of denitrification loss in lowland rice soils. Dinitrogen would appear to be the major gaseous product of denitrification in anaerobic soils because the capacity for reduction of N_2O to N_2 is much greater and also there are more limitations of terminal electron acceptors in anaerobic soils than in the well-aerated or upland soils, and thus N_2O is reduced more rapidly in anoxic soils [30].

The most accepted pathway of denitrification is

$$NO_3^- \rightarrow NO_2^- \rightarrow NO \rightarrow N_2O \rightarrow N_2$$
$$(+5) \quad (+3) \quad (+2) \quad (+1) \quad (0)$$

According to Delwiche [25] considering only the free energy change for the dissimilatory reduction of NO_3^- ion, the most efficient reaction with limited supply of organic substrate is that which results in the production of N_2. He hypothesizes that production of N_2O indicates some reaction barrier involving activation energy of some intermediate products which prevent the full utilization of the energy:

	$-\Delta G'298$ at pH 7	
	(per H_2)	(per NO_3^-)
$NO_3^- + 2H_2 + H^+ \rightarrow 1/2N_2O + 2\ 1/2H_2O$	46.67	93.35
$NO_3^- + 1\ 1/2H_2 + H^+ \rightarrow 1/2N_2 + 3H_2O$	53.62	134.07

where $\Delta G'298$ is the free energy change at the pH indicated.

Recently, Qi and Hua-Kuei [68] reported the isolation of a NO_2^- bacteria from lowland rice soil which oxidizes NH_4^+ to NO_2^- under anaerobic cultural conditions. The organism is a facultative anaerobic ecotype of NO_2^- bacteria and is reported to be closely associated with the denitrifying bacteria. This interesting association of the nitrifying and denitrifying bacteria suggests a unique route of loss of NH_4^+-N to N_2 through NO_2^- at the same location in a flooded soil.

The effect of the presence of plants on denitrification in a flooded soil has not been satisfactorily resolved and reports have indicated both positive and negative results [29, 92].

Dissimilatory nitrate reduction

Thermodynamically, under conditions of abundant organic substrate and limited availability of electron acceptors, the reduction of NO_3^- to NH_4^+ would be more efficient than the formation of N_2 [25].

$$NO_3^- + 4H_2 + 2H^+ \rightarrow NH_4^+ + 3H_2O$$

with $-\Delta G'298$ at pH 7 being 37.25 per H_2 or 149.00 per NO_3^-.

This has been verified by studies that have shown that significant amounts of NO_3^--N may be reduced to NH_4^+-N in anaerobic soils or sediments [10, 11, 14, 43, 45, 48, 98, 100, 103]. The process is termed dissimilatory because it is not inhibited by NH_4^+ or glutamine. These reports indicated that up to 50% of the NO_3^--N could be reduced to NH_4^+-N in some situations, particularly in highly reduced sediments. However, the significance of this process in the N economy of lowland rice soils under field conditions is yet to be ascertained. There is little doubt that if it does occur it could be an important process for conserving N from loss.

Urea hydrolysis

General

Comparatively less emphasis has been placed on urea hydrolysis and urease activity meansurements in flooded soils compared to upland soils. We do know that urease is common in flooded soils. DeLaune and Patrick [24] reported that urease activity was affected by pH but not by water content. Urea hydrolysis occurred in the soil and was negligible in the floodwater overlying the soils. Sahrawat [79] studied urease activity in some Philippine lowland rice soils and found that the urease activity was the lowest in two acid sulfate soils but was higher in mineral soils with near-natural pH and an organic soil. The urease activity in a near-neutral clay was not affected by adding 0.5% NaCl but was markedly increased by 1.3% Na_2CO_3 addition. The floodwater of 11 diverse Philippine lowland rice soils collected from field or greenhouse experiments indicated that the urease activity varied markedly among soils. Except for the floodwater from an acid sulfate soil, all other water samples exhibited significant amounts of urease activity. The highest urease activity was detected in the floodwater of a submerged Histosol. Mineral soils with high pH showed higher urease activity in their

Table 7. Urease activity in floodwater of some Philippine lowland rice soils[a]

| Soil | Floodwater | | Urease activity[b] | |
	Source	pH	Range[c]	Mean
Calalahan sandy loam	Greenhouse	3.9	0–0	0
Quingua silty loam	Greenhouse	6.8	4–6	5
Luisiana clay	Field	6.0	5–7	6
Buenavista clay loam	Greenhouse	7.6	5–7	6
Maahas clay salinized	Field	8.8	6–10	8
Maahas clay	Field	8.0	8–11	9
Pila clay	Greenhouse	7.6	8–11	10
Paete clay loam	Greenhouse	7.5	10–14	12
Lipa loam	Field	8.5	13–19	16
Maahas clay, alkalized	Field	9.4	27–31	29
Lam Aw peat	Field	6.0	33–40	36

[a]From Sahrawat [79].
[b]Urease activity expressed as μg NH_4^+-N formed 25 ml floodwater^{-1} h^{-1} at 30°C.
[c]Four analyses.

floodwaters than those with lower pH (Table 7). This study suggested that the urease activity in floodwater of some soils would hydrolyze significant quantities of surface-applied urea. Studies are needed to evaluate the various factors that affect the urease activity in surface waters because this may be important in relation to NH_3 volatilization loss.

Sahrawat [87] found that the urease activity to ten Philippine lowland rice soils was highly significantly correlated with total N ($r = 0.91^{**}$) and organic C ($r = 0.89^{**}$), but was not significantly correlated with other soil properties. Multiple regression analyses showed that organic matter content of these soils as measured by organic C and total N accounted for most of the variation in urease activity.

Vlek et al. [109] studied urea hydrolysis in three flooded soils and reported that the hydrolysis of urea occurred at the floodwater-soil interface. They showed that the urease activity in the flooded soils was dynamic and was affected by the length of the presubmergence period.

Savant and DeDatta [92] have summarized the recent results obtained on the transformations of different kinds of urea fertilizers in lowland rice soils.

Control of urea hydrolysis

Control of urea hydrolysis in lowland rice soils has received little research attention compared to arable soils [54, 80]. It would be advantageous to control urea hydrolysis in flooded soils since this would decrease N loss due to NH_3 volatilization. However, any advantage may be offset by leaching of urea [54, 80].

Use of controlled release urea-based fertilizers or formulations of urea with urease inhibitors are the two approaches most often suggested for slowing urea hydrolysis. In a recent study, Vlek et al. [109] evaluated the effect of

32

three urease inhibitors with and without an algicide on urea hydrolysis in three flooded soils. Application of potassium ethyl-xanthate and 3-amino-1-H-1,2,4-triazole at 2% (w/w) of urea had no effect on urea hydrolysis or on the dynamics of NH_4^+ concentration in the flood water of soils. Phenylphosphorodiamidate (PPD) [49] applied at 1% (w/w) of urea was very effective in retarding urea hydrolysis for three days. Use of an algicide (simazine herbicide) application to the floodwater of the soils depressed the concentration of NH_3 in floodwater but had little effect in the presence of PPD. A subsequent study by Byrnes et al. [13] showed that PPD was effective in retarding urea hydrolysis in a flooded soil and that its use decreased the loss from 23% to 9% of the ^{15}N applied in a greenhouse experiment. However, application of PPD lowerd dry matter production of rice.

Summary and perspectives

Critical pathways for control

The review of literature on N transformations in flooded soil indicates that NH_3 volatilization could be an important mechanism of loss, especially with urea fertilizers. Control of NH_3 volatilization losses from flooded soils could be achieved by: (i) Placement of the fertilizer in the reduced layer and by proper timing of its application. (ii) Use of algicides may help stabilize pH changes in flood waters and thereby reduce losses of volatile NH_3 [109]. However, use of an algicide may retard biological N_2 fixation in the floodwater and this aspect needs to be carefully evaluated before recommending their use. (iii) Some recent studies have suggested that PPD is an effective blocker of urease activity in flooded soils, and in improving the recovery of fertilizer N by rice under greenhouse conditions. Field studies are needed to further evaluate the efficacy of this and other urease inhibitors for their role in minimizing losses from fertilizer urea. As mentioned earlier, the advantage of retarding urea hydrolysis in some situations may be offset by leaching of urea and should be considered while recommending their use. (iv) Control release of urea can be achieved by use of sulfur-coated ureas and larger granules of urea (called urea supergranules). Their slow-release characteristic combined with the ease in their point application in a flooded soil further increases their efficacy to retard urea hydrolysis and subsequent loss as volative NH_3 [92].

Control of nitrification to control denitrification

Since denitrification occurs only when NO_3^- is present, the best way to control this mechanism of N loss is to minimize nitrification. Also, to date there are no chemicals available that can retard denitrification directly. There is an obvious need to develop chemicals that are cheap and effective inhibitors of nitrification in a flooded soil water system with a wide range in oxidation status. Placement and timing of the fertilizer is probably the most cost-effective means of reducing losses of N due to nitrification-denitrification. Use of urea supergranules or coated fertilizer with controlled release of N which allows plants to compete with microorganisms for fertilizer N should help in reducing N loss by any mechanism, including nitrification-denitrification.

Leaching

In flooded soils with sandy texture, the losses of N due to leaching could be significant. Under these situations, nitrification inhibitors should be more effective than urease inhibitors in minimizing loss of NO_3^-. Urease inhibitors and urea supergranules under these specific high percolation soil conditions may not have any advantage. Perhaps the best answer to minimize leaching loss of N still lies in cultural practices such as split application of fertilizer N and puddling of the rice fields before planting. Slow-release sulfur-coated urea also minimizes N losses by leaching and maximizes N use efficiency.

Need for management

It becomes clear from the foregoing discussion that the most economic and at times even the most effective way of minimizing N losses by controlling critical pathways of N transformations, lies in the best crop and soil management practices. These practices allow the plants to compete effectively with microorganisms involved in the loss of N and thus help in minimizing such losses. Any mechanism or management practice which minimizes NH_4^+ availability for nitrification or volatilization, including plant uptake, should minimize losses of N. We need also a clearer picture of N mineralization patterns so that crop N needs and N release patterns from the soil organic matter can be harmonized.

Acknowledgements

Research supported by the College of Agricultural and Life Sciences, University of Wisconsin-Madison and by the National Science Foundation Grant no. DEB-7817404.

References

1. Abichandan CT and Patnaik S (1958) Nitrogenous fertilizers for rice-ammonium or nitrate form? Rice News Teller 6:6–10
2. Abichandani CT and Patnaik S (1961) Effect of lime application on nitrogen availability and rice yields in waterlogged soils. J Indian Soc Soil Sci 9:55–62
3. Acharya CN (1935) Studies on the anaerobic decomposition of plant materials. III. Comparison of the course of decomposition under anaerobic, aerobic and partially aerobic conditions. Biochem J 29:1116–1120
4. Avnimelech Y and Laher M (1977) Ammonia volatilization from soils: Equilibrium considerations. Soil Sci Soc Am J 41:1080–1084
5. Bajwa MI (1982) Soil clay mineralogies in relation to fertility management:

Effect of clay mineral types on ammonium fixation under conditions of wetland rice culture. Agron J 74:143–144
6. Borthakur HP and Mazundar NN (1968) Effect of lime on nitrogen availability in paddy soil. J Indian Soc Scoil Sci 16:143–147
7. Broadbent FE (1978) Transformations of soil nitrogen. In Nitrogen and rice, pp. 543–559. Los Baños, Philippines: IRRI
8. Broadbent FE (1979) Mineralization of organic nitrogen in paddy soils. In Nitrogen and rice, pp. 105–118. Los Baños, Philippines: IRRI
9. Broadbent FE and Tusneem ME (1971) Losses of nitrogen from some flooded soils in tracer experiments. Soil Sci Soc Am Proc 35:922–926
10. Buresh RJ and Patrick WH Jr (1978) Nitrate reduction to ammonium in anaerobic soil. Soil Sci Soc Am J 42:913–918
11. Buresh RJ and Patrick WH Jr (1981) Nitrate reduction to ammonium and organic nitrogen in an estuarine sediment. Soil Biol Biochem 13:279–283
12. Buresh RJ, Casselman ME and Patrick WH Jr (1980) Nitrogen fixation in flooded soil systems, a review. Adv Agron 33:149–192
13. Byrnes BH, Savant NK and Craswell ET (1983) Effect of a urease inhibitor phenyl phosphorodiamidate on the efficiency of urea applied to rice. Soil Sci Soc Am J 47:270–274
14. Caskey WH and Tiedje JM (1979) Evidence for Clostridia as agents of dissimilatory reduction of nitrate to ammonium in soils. Soil Sci Soc Am J 43:931–936
15. Chandler RF Jr (1979) Rice in the tropics: A guide to the development of National Agencies. Westview Press, Boulder, Colorada 256 p
16. Craswell ET and Vlek PLG (1979) Fate of fertilizer nitrogen applied to wetland rice. In Nitrogen and rice, pp. 175–192. Los Baños, Philippines: IRRI
17. Craswell ET and DeDatta SK (1980) Recent developments in research on nitrogen fertilizers for rice. IRRI Res Pap Ser 49:11 p
18. Craswell ET, DeDatta SK, Obcemea WN and Hartantyo M (1981) Time and mode of nitrogen fertilizer application to tropical wetland rice. Fert Res 2:247–259
19. Daikuhara G and Imasaki T (1907) On the behaviour of nitrate in paddy soils (in Japanese). Bull Imp Cent Agric Expt Stn (Japan) 1:7–37
20. De PK (1936) The problems of the nitrogen supply of rice. I. Fixation of nitrogen in the rice soils under waterlogged conditions. Indian J Agric Sci 6:1237–1245
21. DeDatta SK and Magnaye CP (1969) A survey of the forms and sources of fertilizer nitrogen for flooded rice. Soils Fert 32:103–109
22. DeDatta SK (1981) Principles and practices of rice production. John Wiley & Sons, New York, 618 p
23. DeDatta SK and Kerim MSAAA (1974) Water nitrogen economy of rainfed rice as affected by soil puddling. Soil Sci Soc Am Proc 38:515–518
24. DeLaune RD and Patrick WH Jr (1970) Urea conversion to ammonia in waterlogged soils. Soil Sci Soc Am Proc 34:603–607
25. Delwiche CC (1978) Biological production and utilization of N_2O. Pageoph 116:414–422
26. Delwiche CC (ed.) (1981) Denitrification, nitrification, and nitrous oxide. John Wiley, New York
27. Denmead OT, Freney JR, Simpson Jr (1979) Nitrous oxide emission during denitrification in a flooded field. Soil Sci Soc Am J 43:716–718
28. Dei Y and Yamasaki S (1979) Effect of water and crop management on the nitrogen-supplying capacity of paddy soils. In Nitrogen and rice, pp. 451–463. Los Baños, Philippines: IRRI
29. Firestone MK (1982) Biological denitrification. In Stevenson FJ (ed) Nitrogen in agricultural soils. Agronomy 22:289–326. Am Soc Agron, Madison, Wis
30. Focht DD (1979) Microbial kinetics of nitrogen losses in flooded soils. In Nitrogen and rice, pp. 119–134. Los Baños, Philippines: IRRI
31. Focht DD and Verstraete W (1977) Biochemical ecology of nitrification and denitrification. In Alexander M (ed) Advances Microbiol Ecology, 1:135–214. Plenum Press, New York
32. Freney JR, Denmead OT, Watanabe I and Craswell ET (1981) Ammonia and

nitrous oxide losses following applications of ammonium sulfate to flooded rice. Aust J Agric Res 32:37–45

33. Fried M and Dean LA (1952) A concept concerning the measurement of available soil nutrients. Soil Sci 73:263–271

34. Hauck RD (1979) Methods for studying N transformations in paddy soils: Review and comments. *In* Nitrogen and rice, pp. 73–93. Los Baños, Philippines: IRRI

35. IAEA (1978) Isotope studies on rice fertilization. Joint FAO/IAEA Div of Atomic Energy in Food Agric Tech Rep Ser No 181, IAEA, Vienna

36. IRRI (1974) Annual report for 1973. Los Baños, Philippines: IRRI

37. Ishizuka Y (1965) Nutrient uptake at different stages of growth. *In* The Mineral nutrition of the rice plant, pp. 199–217. Johns Hopkins Press, Baltimore, Maryland

38. Isirimah NO and Keeney DR (1973) Nitrogen transformations in aerobic and water-logged histosols. Soil Sci 115:123–129

39. Jansson SL and Persson J (1982) Mineralization and immobilization of soil nitrogen. *In* Stevenson FJ (ed) Nitrogen in agricultural soils. Agronomy 22:229–252. Am Soc Agron, Madison, Wis

40. Kai H and Wada K (1979) Chemical and biological immobilization of nitrogen in paddy soils. *In* Nitrogen and rice, pp. 157–174. Los Baños, Philippines: IRRI

41. Kelley WP (1911) The assimilation of nitrogen by rice. Hawaii Agric Expt Stn Bull No. 24:20 p

42. Keeney DR (1973) The nitrogen cycle in sediment-water systems. J Environ Qual 2:15–29

43. Keeney DR, Chen RL and Graetz DA (1971) Denitrification and nitrate reduction in sediments: Importance to the nitrogen budget of lakes. Nature 233:66–67

44. Keerthisinghe G, Mengel K and DeDatta SK (1984) The release of nonexchangeable ammonium (^{15}N labelled) in wetland rice soils. Soil Sci Soc Am J 48:291–294

45. Koike J and Hattori A (1978) Denitrification and ammonia formation in anaerobic coastal sediments. Appl Environ Microbiol 35:278–282

46. Koyama T (1981) The transformations and balance of nitrogen in Japanese paddy fields. Fert Res 2:261–278

47. MacRae IC, Ancajas RR and Salandanan S (1968) The fate of nitrate in some tropical soils following submergence. Soil Sci 105:327–334

48. Manguiat IJ and Broadbent FE (1977) ^{15}N studies on nitrogen losses and trans-formations of residual nitrogen in a flooded soil-plant system. Philipp Agric 60:354–366

49. Matzel W, Heber R, Ackermann W and Teske W (1978) Ammoniakverlustre bei Harnstoffdüngung. 3. Beeinflussung der Ammoniakverflüchtignung durch Urease-hemmer. Arch Acker-Pflanzenbau Bodenkunde 22:185–191

50. Mengel K (1982) Factors of plant nutrient availability relevant to soil testing. Plant Soil 64:129–138

51. Mikkelsen DS and DeDatta SK (1979) Ammonia volatilization from wetland rice soils. *In* Nitrogen and rice, pp. 135–156. Los Baños, Philippines: IRRI

52. Mikkelsen DS, DeDatta SK and Obcemae WN (1978) Ammonia volatilization losses from flooded rice soils. Soil Sci Soc Am J 42:725–730

53. Mitsui S (1955) Inorganic nutrition, fertilization and soil amelioration for lowland rice. Yokendo Ltd, Tokyo

54. Mulvaney RL and Bremner JM (1981) Control of urea transformations in soils. *In* Paul EA and Ladd JN (eds) Soil Biochemistry 5:153–196. Marcel Dekker, New York

55. Nelson DW (1982) Gaseous loss of nitrogen other than through denitrification. *In* Stevenson FJ (ed) Nitrogen in agricultural soisl. Agronomy 22:327–363. Am Soc Agron, Madison, Wis

56. Nemeth K (1979) The availability of nutrients in the soil as determined by electro-ultrafiltration (EUF). Adv Agron 31:155–188

57. Nommik H and Vahtras K (1982) Retention and fixation of ammonium and ammonia in soils. *In* Stevenson FJ (ed) Nitrogen in agricultural soils. Agronomy 22:123–171. Am Soc Agron, Madison, Wis

58. Park K, Hood DW and Odum HT (1958) Diurnal pH variations in Texas bays and

its application of primary production estimates. Publ Inst Mar Sci Univ Texas 5:47–64

59. Patrick WH Jr (1982) Nitrogen transformations in submerged soils. *In* Stevenson FJ (ed) Nitrogen in agricultural soils. Agronomy 22:449–465. Am Soc Agron, Madison, Wis

60. Patrick WH Jr and Mahapatra IC (1968) Transformation and availability to rice of nitrogen and phosphorus in waterlogged soils. Adv Agron 20:323–359

61. Patrick WH Jr and Sturgis MB (1955) Concentration and movement of oxygen as related to absorption of ammonium and nitrate nitrogen by rice. Soil Sci Soc Am Proc 19:59–62

62. Payne WJ (1981) Denitrification. Wiley, New York, 214 p

63. Pearsall WH (1950) The investigation of wet soils and its agricultural implications. Emp J Agric 18:289–298

64. Pearsall WH and Mortimer CH (1939) Oxidation-reduction potentials in water-logged soils, natural waters and muds. J Ecol 27:483–501

65. Ponnamperuma FN (1958) Lime as a remedy for a physiological disease of rice associated with excess iron. Int Rice Comm Newsl 7:10–13

66. Ponnamperuma FN (1965) Dynamic aspects of flooded soils. *In* Mineral nutrition of the rice plant, pp. 295–328. John Hopkins Press, Baltimore, Maryland

67. Ponnamperuma FN (1972) The chemistry of submerged soils. Adv Agron 24:29–96

68. Ponnamperuma FN (1978) Electrochemical changes in submerged soils and the growth of rice. *In* Soils and rice, pp. 421–441. Los Baños, Philippines: IRRI

69. Prasad R and DeDatta SK (1979) Increasing fertilizer nitrogen efficiency in wetland rice. *In* Nitrogen and rice, pp. 465–484. Los Baños, Philippines: IRRI

70. Qi Z and Hua-Kuei C (1983) The activity of nitrifying and denitrifying bacteria in paddy soil. Soil Sci 135:31–34

71. Reddy KR (1982) Nitrogen cycling in a flooded-soil ecosystem planted to rice (*Oryza sativa* L.). Plant Soil 67:209–220

72. Reddy KR, Patrick WH Jr and Phillips RE (1976) Ammonium diffusion as a factor in nitrogen loss from flooded soil. Soil Sci Soc Am J 40:528–533

73. Reddy KR, Patrick WH Jr and Phillips RE (1978) The role of nitrate diffusion in determining the order and rate of denitrification in flooded soil: I. Experimental results. Soil Sci Soc Am J 42:268–272

74. Reddy KR, Patrick WH Jr and Phillips RE (1980) Evaluation of selected processes controlling nitrogen loss in flooded soil. Soil Sci Soc Am J 44:1241–1246

75. Sahrawat KL (1979a) Nitrogen losses in rice soils. Fert News 24:38–48

76. Sahrawat KL (1979b) Evaluation of some chemical extractants for determination of exchangeable ammonium in tropical rice soils. Commun Soil Sci Plant Anal 10:1005–1013

77. Sahrawat KL (1979c) Ammonium fixation in some tropical rice soils. Commun Soil Sci Plant Anal 10:1015–1023

78. Sahrawat KL (1980a) Nitrogen mineralization in acid sulfate soils. Plant Soil 57:143–146

79. Sahrawat KL (1980b) Urease activity in tropical rice soils and flood water. Soil Biol Biochem 12:195–196

80. Sahrawat KL (1980c) Control of urea hydrolysis and nitrification in soil by chemicals-prospects and problems. Plant Soil 57:335–352

81. Sahrawat KL (1981) Ammonification in air-dried tropical lowland histosols. Soil Biol Biochem 13:323–324

82. Sahrawat KL (1982a) Evaluation of some chemical indexes for predicting minerali-zable nitrogen in tropical rice soils. Commun Soil Sci Plant Anal 13:363–377

83. Sahrawat KL (1982b) Simple modification of the Walkley-Black method for simultaneous determination of organic carbon and potentially mineralizable nitrogen in tropical rice soils. Plant Soil 69:73–77

84. Sahrawat KL (1982c) Nitrification in some tropical soils. Plant Soil 65:281–286

85. Sahrawat KL (1983a) Mineralization of soil organic nitrogen under waterlogged conditions in relation to other properties of tropical rice soils. Aust J Soil Res 21:133–138

86. Sahrawat KL (1983b) Nitrogen availability indexes for submerged rice soils. Adv Agron 36:415–451
87. Sahrawat KL (1983c) Relationships between soil urease activity and other properties of some tropical wetland rice soils. Fert Res 4:145–150
88. Sahrawat KL and Burford JR (1982) Modification of the alkaline permanganate method for assessing the availability of soil nitrogen in upland soils. Soil Sci 133:53–57
89. Sanchez PA (1972) Nitrogen fertilization and management in tropical rice. North Carolina Agric Expt Stn Tech Bull 213:1–31
90. Sanchez PA (1973) Puddling tropical rice soils. II. Effect of water losses. Soil Sci 115:303–308
91. Savant NK and DeDatta SK (1980) Movement and distribution of ammonium-N following deep placement of urea in a wetland rice soil. Soil Sci Soc Am J 44:559–565
92. Savant NK and DeDatta SK (1982) Nitrogen transformations in wetland rice soils. Adv Agron 35:241–302
93. Sethunathan N, Rao VR, Adhya TK and Raghu K (1983) Microbiology of rice soils. CRC Critic Rev Microbiol 10:125–172
94. Shiga H and Ventura W (1976) Nitrogen supplying ability of paddy soils under field conditions in the Philippines. Soil Sci Plant Nutr 22:387–399
95. Shioiri M (1948) Effect of drying in paddy fields during fallow period. Bull Agric Expt Stn Minist Agric (Japan) 64:1–24
96. Shioiri M, Aomine S, Uno Y and Harada T (1941) Effect of air-drying of paddy soil (in Japanese). J Sci Soil Manure (Japan) 15:331–333
97. Smith CJ, Brandon M and Patrick WH Jr (1982) Nitrous oxide emission following urea-N fertilization of wetland rice. Soil Sci Plant Nutr 28:161–171
98. Sorensen J (1978) Capacity for denitrification and reduction of nitrate to ammonia in a coastal marine sediment. Appl Environ Microbiol 35:301–305
99. Stanford G (1978) Evaluation of ammonium release by alkaline permanganate extraction as an index of soil nitrogen availability. Soil Sci 126:244–253
100. Stanford G, Legg JO, Dzienia S and Simpson EC Jr (1975) Denitrification and associated nitrogen transformations in soils. Soil Sci 120:147–152
101. Tanji KK (1982) Modeling of the soil nitrogen cycle. In Stevenson FJ (ed) Nitrogen in agricultural soils. Agronomy 22:721–772. Am Soc Agron, Madison, Wis
102. Terman GL (1979) Volatilization losses of nitrogen as ammonia from surface-applied fertilizers, organic amendments, and crop residues. Adv Agron 31:189–223
103. Tiedje JM, Firestone RB, Firestone MK, Betlach MR, Kaspar HF and Sorensen J (1981) Use of nitrogen-13 in studies of denitrification. In Krohn KA and Root JW (eds) Recent developments in biological and chemical research with short-lived isotopes. Advances in Chemistry, Am Chem Soc, Washington
104. Tilo SN, Caramoncion MDV, Manguiat IJ and Paterno ES (1977) Naturally occurring fixed ammonium in some Philippine soils. Philipp Agric 60:413–419
105. Tusneem ME and Patrick WH Jr (1971) Nitrogen transformations in waterlogged soil. Louisiana State Univ Agric Expt Stn Bull 657:1–75
106. Ventura W and Watanabe I (1978) Dry season soil conditions and soil nitrogen availability to wet season wetland rice. Soil Sci Plant Nutr 24:533–545
107. Vlek PLG, Byrnes BH and Craswell ET (1980) Effect of urea placement on leaching losses of nitrogen from flooded rice soils. Plant Soil 54:441–449
108. Vlek PLG and Craswell ET (1981) Ammonia volatilization from flooded soils. Fert Res 2:227–245
109. Vlek PLG, Stumpe JM and Byrnes BJ (1980) Urease activity and inhibition in flooded soil systems. Fert Res 1:191–202
110. Watanabe I and Cholitkul W (1979) Field studies on nitrogen fixation in paddy soils. In Nitrogen and rice, pp. 223–239. Los Baños, Philippines: IRRI
111. Williams WA, Mikkelsen DS, Mueller KE and Ruckman JR (1968) Nitrogen immobilization by rice straw incorporated in lowland rice production. Plant Soil 28:49–60

112. Yoshida T (1975) Microbial metabolism of flooded soils. *In* Paul EA and McLaren AD (eds) Soil Biochemistry 3:83–122
113. Yoshida T (1978) Microbial metabolism in rice soils. *In* Soils and rice, pp. 445–463. Los Baños, Philippines: IRRI
114. Yoshida T and Ancajas RR (1971) Nitrogen fixation by bacteria in the root zone of rice. Soil Sci Soc Am Proc 35:156–157
115. Yoshida T and Ancajas RR (1973) Nitrogen-fixing activity in upland and flooded rice fields. Soil Sci Soc Am Proc 37:42–46
116. Yoshida T and Padre BC Jr (1974) Nitrification and denitrification in submerged Maahas clay soil. Soil Sci Plant Nutr 20:241–247

3. Technologies for utilizing biological nitrogen fixation in wetland rice: potentialities, current usage, and limiting factors

PA ROGER* and I WATANABE

The International Rice Research Institute, P.O. Box 933, Manila, Philippines

Key words: nitrogen fixation, lowland rice, azolla, blue-green algae, legumes, straw, technology, review

Abstract. Almost all types of N_2-fixing microorganisms are found in lowland rice fields. The resulting N fertility has permitted moderate but constant productivity in fields where no N fertilizer is applied. Current and potential technologies for utilizing biological N_2 fixation in lowland rice production are reviewed in terms of potential, current usage, and limiting factors.

Legumes and azolla have been traditionally used as green manure in parts of Asia, permitting yields of 2–4 t/ha. To a limited extent, straw incorporation favors heterotrophic biological N_2 fixation. Recently, inoculation with blue-green algae has been claimed to increase yields by about 10%. Using non-symbiotic N_2-fixing systems is still experimental.

Utilization of biological N_2 fixation as an alternative or additional N source for rice is severely limited by technological, environmental, and socioeconomical factors.

Rice is the staple food of approximately one half of the world's people. About 75% of the 143 million hectares of ricelands are lowlands where rice grows in flooded fields during all or part of the cropping period.

Lowland rice can be grown on the same land year after year without N fertilizer and produce moderate but constant yields. In contrast, upland rice yields decline over time if no N fertilizer is applied. The continuing N fertility in lowland rice fields has been attributed to higher N_2 fixing activity coupled with slower decomposition of organic N compounds under poorly aerated conditions [15].

N is usually the limiting factor to produce high yields. The green revolution in rice production is based on fertilizer--responsive rice varieties. In Asia, one of the constraints to high yields is the limited availability and high prices of N and P fertilizers. In 1978 fertilizer use in tropical Asian countries averaged 30–55 kg NPK/ha arable land [92]. The idea of utilizing biological N_2 fixation (BNF) as an alternative or supplementary source of N for rice is not new. N_2-fixing green manures have been used for centuries in some rice growing areas, and research on biofertilizers, including algal and bacterial inoculants, began in the early 1930s.

*Maitre de Recherches ORSTOM (France). Visiting Scientist at IRRI.

Fertilizer Research 9 (1986) 39–77
© *Martinus Nijhoff/Dr W. Junk Publishers, Dordrecht – Printed in the Netherlands*

Biological N_2 fixation in flooded soil systems was reviewed by Buresh et al. [15]. Processes and ecology of BNF in rice soils were reviewed by Watanabe [147] and Watanabe and Brotonegoro [152]. An extensive study of quantitative data was presented in Lowendorf's review [75]. New knowledge of BNF in flooded rice fields was summarized by Watanabe and Roger [159].

This paper intends to describe the potential for practical utilization of BNF technology by rice farmers rather than to discuss mechanisms and N_2^- fixing agents. After a short summary of the properties of the lowland rice field ecosystem as a site for BNF, emphasis is placed on technologies, their potential and current usage, and the technological, environmental, and socio-economical factors that limit adoption. Although most legume green manures are grown in upland conditions before lowland rice, we found it appropriate to include legume green manuring in our discussion. Straw incorporation, seldom considered in previous reviews, is also discussed.

I The wetland rice field ecosystem as a site for BNF

A Agronomic characteristics of lowland rice fields

Wetlands (lowlands) were defined by Cowardin et al. [23] as 'lands where saturation with water is the dominant factor determining the nature of soil development and the types of plant and animal communities. Flooding the soil has many advantages for rice production:
– it provides a continuous water supply to the crop;
– it changes the pH of alkaline and acidic soils towards neutrality or slight acidity which is favorable for rice growth;
– it diminishes the incidence of soil sickness and outbreak of soil borne diseases usually observed under continuous monocropping in upland soils;
– it depresses weed growth, especially C_4 type grasses;
– it favors BNF, giving flooded soils a higher spontaneous fertility than upland soils;
– irrigation water supplies nutrients such as Ca, Si, and K; and
– bunded rice fields act as water reservoirs and prevent soil erosion.

B Different sites for BNF in rice fields

Principal environmental characteristics of lowland rice fields are determined by: flooding, the presence of rice plants, and agricultural practices.

Flooding of the soil soon creates anaerobic conditions in the reduced layer, a few millimeter beneath the soil surface. Flooding and rice plants lead to the differentiation of five major environments differing by their physico-chemical and trophic properties and the energy sources for BNF: floodwater, surface oxidized soil, reduced soil, rice plants (submerged parts and rhizo-sphere), and subsoil (Fig. 1).

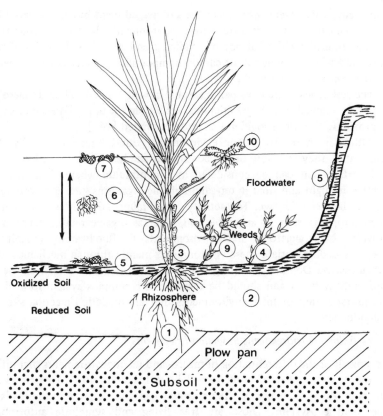

Figure 1. Diagram of environments and N_2-fixing components in a rice field ecosystem. *N_2 fixing bacteria:* 1) associated with the roots, 2) in the soil, 3) epiphytic on rice, 4) epiphytic on weeds. *blue-green algae:* 5) at soil-water interface, 6) free floating, 7) at air-water interface, 8) epiphytic on rice, 9) epiphytic on weeds. *azolla:* 10

The floodwater is a photic, aerobic environment where aquatic communities of producers (algae and aquatic weeds) and consumers (bacteria, zooplankton, invertebrates, etc.) provide organic matter to the soil and recycle nutrients.

The oxidized soil layer is a photic aerobic environment with a positive redox potential [161], a few millimetres thick, where NO_3^-, Fe^{+3}, SO_4^{-2}, and CO_2 are stable and where algae and aerobic bacteria predominate.

The reduced soil layer is nonphotic anaerobic environment where Eh is predominantly negative [161], reduction processes predominate, and where microbial activity is concentrated in soil aggregates containing organic debris [145].

The rice plant comprises two major subenvironments: submerged plant parts and the rhizosphere. In floodwater, basal portions of rice shoots and aquatic weeds are colonized by epiphytic bacteria and algae. Epiphytism is important in deepwater rice where the submerged plant biomass is very high [115]. The rhizosphere is a nonphotic environment where redox conditions are determined by the balance of oxidizing and reducing capacities of rice roots and where production of carbon compounds by roots provides energy sources for microbial growth.

The soil beneath the plow pan is aerobic in well-drained soils and anaerobic in poorly drained soils. It has microbial activity in the upper layer and its role in providing N to rice should not be underestimated [143].

Although those five major environments can be macroscopically differentiated, they are more or less continuous and heterogeneous. Floodwater and surface soil can be considered as a continuum where algae become benthic in night and float in day and epipelic organisms such as chironomids migrate between the two. The activity of soil fauna induces the formation of microaerophilic sites in anaerobic soil, whereas organic matter debris may provide some anaerobic microenvironments in floodwater. Agricultural practices dedifferentiate these environments mainly through mechanical disturbances. As Dommergues [32] wrote, flooded rice soil is far from being uniformly reduced and should be regarded as a complex system formed by the juxtaposition of microenvironments that are oxidation reaction sites or reduction sites.

C N_2-fixing microorganisms in wetland rice fields

As a result of the differentiation of macro- and microenvironments that differ by redox state, physical properties, light status, and nutritional sources for the microflora, all major N_2-fixing groups can and do grow in the lowland rice field ecosystem. Those are free living and symbiotic autotrophs, symbiotic heterotrophs, and aerobic, facultative anaerobic, and anaerobic free living heterotrophs (Table 1). The floodwater, the submerged plant biomass, and the aerobic soil layer are sites of photodependent N_2 fixation. Heterotrophic N_2 fixation develops preferentially in nonphotic environments: the soil aggregates that contain organic debris and the rhizospere.

From an ecological point of view, N_2-fixing organisms in rice fields can be classified as:
— three groups of autotrophs comprising photosynthetic bacteria, free living blue-green algae (BGA), and azolla, and
— three groups of heterotrophs comprising N_2-fixing bacteria in the soil, N_2-fixing bacteria associated with rice, and legume green manures.

D Evidence and quantification of BNF in lowland rice fields

A summary of data on N uptake by rice crops in nonnitrogen plots of long-term fertility trials [154] indicates that in the absence of N fertilizer, a rice crop uses an average 40–50 kg N/ha. In most of those experiments soil N did not decrease, indicating that used N is compensated for by mechanisms among which BNF is most important.

Table 1. Major groups of N₂-fixing microorganisms in lowland rice fields

Photoautotrophs	Free living		Photosynthetic bacteria	*Rhodopseudomonas*
			Blue-green algae (cyanobacteria)	*Nostoc, Anabaena*
	Symbiotic			*Anabaena azollae* in *Azolla* sp.
Heterotrophs	Free living	In the soil		
			Oxidized soil:	
			obligate aerobes*	*Azotobacter, Beijerinckia*
			microaerobes**	*Methylomonas*
			Reduced soil:	
			obligate anaerobes	*Clostridium, Desulfovibrio*
		In association with rice		
			On submerged parts:	
			facultative anaerobes*	*Klebsiella*
				Enterobacter
			In the rhizosphere:	
			facultative anaerobes**	*Flavobacterium, Pseudomonas*
			microaerobes**	*Azospirillum*
	Symbiotic			*Rhizobium* in legumes

*Growth in presence of O₂ but fix N only in absence of O₂
**Fix only in the presence of low O₂ concentrations

Quantification of BNF during the crop cycle has been tried, unfortunately the methods were inaccurate and controversial. Nitrogen balance experiments give only the sum of N gains and losses. In field experiments the contribution of subsoil is generally not estimated [154], and in pot experiments the contribution of phototrophs may be overestimated [115]. The limitations of the acetylene reduction assay and ^{15}N methods were summarized by Watanabe and Roger [159].

Long term fertility experiments show that N balance between losses and inputs through BNF and other minor N sources range from 20 to 70 kg N/ha and per crop in plots receiving no N fertilizer [154]. Evaluation of photo-dependent BNF, compiled by Roger and Kulasooriya [115], ranged from very little to 80 kg N/ha per crop and averaged 27 kg N/ha per crop. An extensive compilation of quantitative estimates of BNF in lowland rice fields is in Lowendorf's [75] review. He summarized the data as ranges of fixed N/ha per crop as: legumes, 25–61 kg/ha; BGA, 0.2–39; azolla green manure, 25–121; azolla dual crop, 2–75; and soil and rhizosphere fixers, 1.2–18.3.

E Available and potential technologies

Potential for a significant agronomical N contribution has been proved for five of the six major groups of N_2-fixing organisms in lowland rice fields. The potential of photosynthetic bacteria has not been assessed. Among those five groups, only N_2-fixing legume and azolla green manures are being used purposefully as a N source for rice. Heterotrophic N_2 fixers are favored by straw incorporation but that technology is not used to increase BNF. A technology for rice fields inoculation with BGA has been developed in some countries but is not used by farmers to a noticeable extent. Technology to enhance the activity of heterotrophs associated with rice roots is unavailable.

II Technologies used by farmers

A Legume green manures

The potential of legume green manures for rice was early recognized. In 1936 the International Institute of Agriculture [49] reported that 'application of green manure may involve great progress in rice growing by ensuring yields higher than those at present attained'. Similar statements were recorded by Pandey and Morris [93] from the proceedings of international meetings in 1952, 1953, and 1954. Since then, less attention has been given to legumes in rice production than to other sources of BNF. Lowendorf [75] wrote that legumes were mentioned in only a few paragraphs of the proceedings of the International Rice Research Institute (IRRI) symposium on *Nitrogen and Rice* [69, 94]. However, emphasis on legumes was increased at the recent IRRI symposium on *Organic Matter and Rice* [66, 106, 126, 139, 160].

1 Potentialities. Based on reported N contents in legume green manure crops, (Table 2) it appears that one crop accumulates on average 100 kg N/ha. Highest values have been reported for *Sesbania* spp: 146 kg N/crop for *S. sireceda* [40], 202 kg for *S. sesban* [51], and 267 kg in 52 d for *S. rostrata* [114]. *Sesbania rostrata* forms N_2-fixing nodules on the roots and the stems, and has 5–10 times more nodules than most legumes. Because of its stem nodules, it can fix N under waterlogged conditions and when the N content of the medium is high, giving it exceptionally high potential for BNF [34, 35]. Assuming that between 50 and 80% of N accumulated in legumes originates from BNF [39] it appears that legume green manures could provide 50–80 kg N to a rice crop.

Because N availability to rice may vary with the kind of green manure, a better evaluation of green manure potential is to compare them with N fertilizers. Incorporating one legume crop is equivalent to applying 30 to 80 kg fertilizer N (Table 3).

Table 2. Legume green manures as a N source. [Data from 11, 40, 51, 52, 149].

Species	N content of a crop (kg/ha)	(% fresh weight)
Astragalus sinicus	108–123	0.35–0.47
Canavalia ensiformis	98	0.47
Cassia mimosoides	97	0.44
Crotalaria anagyroides	98	0.33
Crotalaria juncea	129	0.30
Crotalaria juncea	105	–
Crotalaria quinquefolia	88	0.19
Dolichos biflorus	89	0.58
Gycine koidzumii	71	0.42
Phaseolus sp.	–	0.28
Phaseolus lathyroid	90	–
Phaseolus calcaratus	42	0.22
Sesbania aculeata	122	0.32
Sesbania aculeata	96	0.36
Sesbania rostrata	267	–
Sesbania sesban	100	0.39
Sesbania sesban	202	–
Sesbania microcarpa	87	0.50
Sesbania sirececa	146	–
Average value	114	0.37
c.v. %	43	29

The effects of incorporating a legume green manure on rice yield and soil properties have been thoroughly documented by field experiments in many countries [93]. The relative efficiency of different legume species and comparisons with chemical fertilizers were discussed by Singh [126]. In China, which has a long history of green manuring, Wen Qi-Xiao [160] reported that on average, applying 1 ton (fresh weight) of winter green

Table 3. Experiments comparing legume green manures and N fertilizers

Incorporated green manure (1 crop)	N fertilizer needed for the same yield (kg/ha)	Reference
Crotalaria juncea	75	11
Sesbania aculeata	50	11
Sesbania (leaves)	40	20
Sesbania (20 kg N)	30	136
Sesbania aculeata	60	10
Sesbania aculeata	65–70	133
Vigna radiata	80	87
Green gram	60	85
Daincha	75	26
1 crop green manure (average)	30–80	93

manure to rice will increase yield by 30–80 kg, depending on soil fertility, rate of application, and rice cultivar. The N utilization rate was estimated to be about 30%. Pandey and Morris [93] estimated legumes potentialities to be 100 kg grain yield increase per ton of green manure incorporated (winter, spring, and summer green manures). Rinaudo et al. [114] reported a record yield increase from *S. rostrata*. Incorporated as a 52-d-old crop before transplanting, it increased rice yield by 3.7 t/ha over the control. Applying 60 kg N as ammonium sulfate increased yield by 1.7 t/ha over the control. About 33% of the N accumulated by *S. rostrata* was transferred to the crop.

Tiwari et al. [133] reported that wheat crop following a rice crop with green manure registered a 54% increase in grain yield compared with the wheat crop that followed rice with no green manuring, indicating residual effects.

Besides increasing rice yield and N content of the grain [133], incorporating legume green manures is thought to have additional beneficial effects such as increasing soil N and organic matter content [122], available Zn [48], hydraulic conductivity, water holding capacity, and aggregate stability of the soil [12]. Another possible beneficial effect is trying up mineralized soil N, thus preventing its loss by denitrification, volatilization, or leaching when land is fallow [75].

2 Methods of utilization. In areas where rainfall is the primary water source in wet season, legumes may be intersown with rice shortly before harvest, grown through dry season, and incorporated when soil is prepared for the next crop [16, 43]. Although those legumes must be drought tolerant, the economic value of legume green manures is less affected by drought than that of grain legumes because green manure is not strongly dependent on drought sensitive reproduction processes.

Where two rice crops are grown or where rainy season is long enough, a green manure can be grown just before rice and incorporated before transplanting. In North Vietnam, *Sesbania* is planted about 60 d before harvesting spring rice and is incorporated, with rice straw, $2\frac{1}{2}$ mo later before the summer rice crop. *Sesbania* biomass is 5–7 t/ha, fresh weight, equivalent to 25–35 kg N. For spring rice, farmers prefer azolla, which supplies more N [50].

Leaves and cuttings of perennial plants grown along field borders, and even wild legumes collected nearby have been used as green manure and applied at or after planting [126]. Devoting fields to green manure production is uncommon. However, Chari [16] wrote that 1 ha of *Sesbania* could provide 10 ha of rice with 6.7–9.0 t of green manure or 40–50 kg N/ha. Growing a cash crop that is harvested for grain and then incorporated into the soil is another green manure alternative [139]. However, under such conditions, soil N may be depleted if more N is removed in the harvest than was fixed [75].

Legume utilization may be improved in integrated management as reported by Wen Qi-Xiao [160]. Using 1,250 kg (fresh weight) of milk vetch (*Astragalus sinicus*) as swine fodder increased the swine's weight by 26 kg. Applying the resulting excreta to rice as fertilizer increased grain yield by 27 kg. Applying milk vetch directly similarly increased rice yield. In that combination, 2-step green manure utilization was more profitable.

3 Present usage Despite their potential to increase yield, use of leguminous green manures has decreased in recent years.

China is the only country where legumes are still widely used. In a review of utilization of organic materials in rice production in the tropical and subtropical zone south of the Huai River (the center of rice production in China), Wen Qi-Xiao [160] wrote that the total area of leguminous green manure crops and azolla in that region had increased from 2.45 million ha in 1952 to 8.35 million ha in 1979 representing an additional 346 000 t N. In 1980, Chen [21] reported that green manure was planted on 20- 50% of the total farmland in Jiangsu, Zhejiang, Jiangxi, Huuan, and Husei provinces. Total green manured area was estimated to be 10 million ha. Although total green manure used for rice production in China has increased since 1949, their relative contribution to N fertilization decreased [160]. The higher the level of intensive agriculture the lower is the ratio of organic manure to chemical fertilizers and the higher the proportion of farm yard manure in organic manure. Hectarage of green manure tends to increase where soil fertility is low and decreases where soil fertility is high.

In other countries, green manure use seems to have become incidental. About 100,000 ha of *S. sesban* are estimated to be grown annually for summer rice in northern Vietnam, when azolla cannot be used because of high pest incidence and high temperatures [50].

In India, Singh [126] reported that 20-30% of several rice growing areas were planted to legume green manure at mid century but that recently green manuring has decreased with increased cropping intensity and ready fertilizer availability. Although green manuring and green leaf manuring are well-known to farmers, they have not been extensive in India since organic fertilizers were introduced [139].

Since 1955 a spectacular decline in green manuring has also been observed in Japan [149].

4 Limiting factors. Many reasons have been given to explain the decline and the non acceptance of green manuring by farmers in South and Southeast Asia.

In temperate climates some detrimental effects of green manuring have been reported. Watanabe [149] attributed the decline of green manures in Japan to:
— possible rice growth reduction caused by anaerobic decomposition of green manure;
— lack of synchronization between N release and plant N needs which depresses growth at early stages and causes excessive growth, detrimental to yield at later stages;
— soil degradation, and;
unpredictability of the amount of N applied.

An important limiting factor is the bulkiness of green manures and resulting incorporation difficulties. Nitrogen content in legumes varies from 0.2 to 0.6% (Table 2) therefore the fresh weight corresponding to 50 kg N/ha varies from 10 to 26 t. Large scale incorporation of such a large and more or less lignified biomass using animal draft power and traditional implements is difficult. In developed countries high wage rates and lacking manpower may be limiting. One of the reasons for the abandonment of organic manure in Japan is the decreasing labor available for farming in combination with the fact that about 90% of rice farmers are part-time farmers [66].

There are other socioeconomic limitations. Legume green manures are not appealing because they do not directly yield food or cash. Also, green manures are poorly competitive or not competitive where commercial N fertilizer is available. Assuming an average yield of 15 kg grain/kg applied N, the cost of inorganic N fertilizer relative to rice price is very favorable [163]. Furthermore, many governments have a fertilizer subsidy policy and made cheap credit available for farmers to buy N fertilizer. The cost of green manure seed and necessary land preparation are unfavorable, given the limited yield response obtained from the green manure [93]. If there is residual soil moisture after rice harvest, farmers prefer to grow food legumes, peanut, maize, millet, onion etc. and apply to inorganic fertilizers for the next rice crop [139]. Economics often favor a food cash crop over green manure.

If N fertilizer is unavailable it is most frequently to subsistence farmers, with small land holdings, who cannot afford to release land to green manure production and prefer growing a food crop. In some areas where no N fertilizer is available and organic manure was traditionally applied to rice, green manure is now applied to vegetable cash crops rather than to rice [134].

There are also incidental limitations. In certain areas of India indiscriminate cattle grazing because of inadequate social control makes farmers reluctant to grow a green manure crop [139].

5 Conclusion. In terms of fixed N, legumes have tremendous potential and there is a large range of drought or submergence adapted green manure species. Within the aquatic legumes, stem-nodulated species (*S. rostrata, Aeschynomene indica, Neptunia oleracea*) represent a further step in adaptation to waterlogging and nitrogen fixation in soil with high level of combined nitrogen [42]. Despite this potential, green manuring is not widespread and has even declined in recent years. As indicated by Pandey and Morris [93] that trend should evoke cautious and thoughtful research.

The same authors concluded that green manuring has realistic potential where the subsistence component of the farm-household complex is high. Such areas are frequently rainfed with no available N fertilizer. They wrote that BNF by a green manure, and conservation of NO_3^- mineralized in dry season may be economically viable if production costs can be kept low, and if the green manure does not compete with marketable or subsistence crops. This finding indicates two possibilities – one is to grow a drought tolerant green manure early in dry season – the other is to grow, early in rainy season, a fast-growing excess-water tolerant green manure species.

Because legume green manures must be cultivated under very different agroclimatic conditions, selection should be done for diverse environmental tolerance, rapid growth, spontaneous nodulation, and high N_2-fixing ability. Additionally, technologies should be developed for specific crop cycles. For example, optimum time for incorporation before flooding is required to avoid denitrification losses. However, this time must also coordinate N demand of rice with green manure N release, which varies by the nature and state of green manure. Therefore, factors such as green manure species, C:N, lignin content, age at incorporation, and delay between incorporation and planting, must be considered.

As pointed out by Morris and Pandey [87] total reliance on organic fertilizers is unrealistic when considering modern, high yielding, N responsive rice varieties. Inorganic N must be applied to obtain desirable high yields. Under such conditions the inorganic N component is of major significance because application timing is more flexible.

It appears that despite a high N_2-fixing potential, green manure usage, known to many farmers, has been abandoned for various reasons. However, green manuring with fast growing, naturally nodulating species, adapted to specific environmental conditions may be important in some cropping patterns. Therefore, considering their present usage, legumes may be considered as an underutilized N source in rice production.

B Azolla

Azolla is a symbiosis between an aquatic fern of the *Azolla genus* and a N_2-fixing blue-green alga, *Anabaena azollae*. Because of its rapid growth and high N content azolla has been used as green manure in rice culture for centuries in northern Vietnam and southern China. Utilization of azolla in those areas was generally ignored by scientists in other countries until the mid-1970s. Since then information gathered by visitors to and from China and Vietnam, articles and reviews [76, 86, 148] have stimulated interest and research on azolla. A book on azolla as a green manure was written in 1982 by Lumpkin and Plucknett [77] and a Primer on azolla by Khan in 1983 [67]. Besides its utilization as green manure, azolla also is used as swine and poultry feed [74]

1 Potential. Growth rate, maximum biomass, N_2-fixing activity, and rice yields in experiments comparing azolla utilization with fertilizer application estimate the potential of azolla for rice production. Growth rates were summarized by Becking [9]. He showed that doubling time varies between 2 and 10 d for most species. Growth follows an approximate logistic curve until maximum biomass is reached. Growth rate declines as plant density increases [4]. Maximum biomasses and N_2-fixing rates were summarized by Kikuchi et al. [68]. Maximum biomasses ranged from 0.8 to 5.2 t dry matter/ha and averaged 2.1 t dry matter/ha and averaged 2.1 t/ha (n = 13; c.v. = 57%). N content ranged from 20 to 146 kg N/ha and averaged 70 kg N/ha (n = 17; c.v. = 58%). N_2-fixing rate ranged from 0.4 to 3.6 kg N/ha per day and averaged 2 kg N/ha per day (n = 15, c.v. = 47%).

In 1 yr, N fixed by continuous azolla culture, repeatedly harvested, can be as high as 500 (*A. pinnata*) [151] and 1200 kg N/ha (*A. filiculoides*) [72]. When grown dual cropped with rice, azolla can accumulate from 25 to 170 kg N/ha in 60 d [68] which may exceed the N requirement of rice. Azolla N_2-fixing activity per unit area is similar or higher than that of legume pastures.

International azolla field trials conducted for 4 consecutive years in 19 sites and 9 countries by the International Network on Soil Fertility and Fertilizer Evaluation for Rice (INSFFER) network [53, 54, 55, 57] have shown that:

1   Incorporating one crop of azolla grown before or after transplanting was equivalent to split application of 30 kg fertilizer N.
2   Incorporating two crops of azolla grown before and after transplanting was equivalent to split application of 60 kg fertilizer N.

3   Increase in rice yield per unit weight of azolla incorporated was roughly proportional to the response of rice to chemical fertilizer at the same site.

4   Spacing of rice plants (20 × 20 cm vs 10 × 40 cm) did not significantly affect azolla growth and rice yield.

2 Methods of utilization. China and Vietnam are the only countries with a long history of azolla cultivation extending back to the 11th century in Vietnam [25] and at least to the Ming dynasty (1368–1644 A.D.) in China [77].

In China, azolla is used from 37 °N (Shandong) to 19 °N (Hainan). Because strains used grow best at a 25 °C average daily temperature, azolla usually is grown in late spring (May to June) in the north and early spring (March to April) in the south. Most frequently it is grown for about 1 mo, then incorporated before transplanting. To a lower extent, the wide-narrow row spacing for transplanted rice [74] is used to grow azolla with rice, which permits several incorporations during the crop cycle. In some places, azolla is grown before and after transplanting.

In Vietnam [50] azolla is used in the northern provinces where spring and summer rice are grown with adequate irrigation. Azolla only is used for spring rice because in summer it grows poorly and insect incidence is severe. Straw incorporation and legume green manuring are used for summer rice.

About 2–3 mo before transplanting, azolla is collected from natural environments by specialists, and is multiplied in government azolla farms at regional and district levels. Inoculum is sold to cooperatives and farmers who propagate it in the fields from November to February, when conditions are favorable for its multiplication. As in China, azolla production and utilization is done two ways; as green manure incorporated before transplanting, and as an intercrop incorporated after transplanting. Azolla grown before transplanting is fertilized with P, K, and farmyard manure. Recommended applications are: 2.2 kg P/ha every 5 d, 4 kg K/ha every 10 d, and 500–1000 kg/ha farmyard manure every 5–10 d. When chemical fertilizers are unavailable, ash is substituted. Intercropped azolla usually is not fertilized but if superphosphate is available one application of 4.5 kg P/ha per crop is recommended.

Azolla is inoculated in the field at 300–500 kg fresh weight/ha with farmyard manure. To facilitate vegetative multiplication, fronds are broken using a special tool. Azolla is collected before heavy rains to protect it from being washed from the fields. When attacked by insects, azolla is collected) and placed under water for 15 h to kill insect larvae. Sixteen to 20 d after inoculation the field is covered with about 20 t of azolla, half of which is collected in mounds and composted by covering it with soil. The remaining half is grown for 7–10 more days after which the field is again fully covered. Half of the second crop and the compost are incorporated in the soil. Rice is

52

transplanted and the remaining half of the second azolla crop keeps on growing. Seven to ten days after transplanting, the field is covered and again half of the third crop is hand-and-foot incorporated between the rows. Sometimes a 4th crop is grown and incorporated.

This technology produces, on average, 40 t/ha per rice crop (fresh weight) of azolla, equalling 80 kg N/ha. It requires application of 0.5 t fresh azolla inoculum, 2–3 t farmyard manure, 20–30 kg P, and | 20 kg K/ha. Average annual rice yield in the Red River Delta is 5–7 t. The winter crop, with azolla, yields 3 to 5 t/ha, and the summer crop, without azolla, yields 2 t or less/ha.

The azolla technology used in Vietnam is labor-intensive and might be improved by changing the transplanting method, thus permitting easy azolla incorporation with a rotary weeder.

3 Current usage. Data concerning the extent of azolla use in China are controversial. FAO [38] estimated that azolla is grown in over 6.5 million ha of rice lands in China. In contrast, Liu Chung-Chu [74] reported that 1.34 million ha were planted to azolla in 1979, which was slightly less than 5% of the total rice area. According to Liu Chung-Chu, azolla is used in rice-growing regions from the South Yantze River to northeast China (40 ˇN) and in the Hanchung Basin of Shensi in the west. Data reported by Lumpkinand Plucknett [77] indicate that azolla is cultivated as a green manure on only 2% of China's harvested rice area and for about 5% of its spring rice crop.

In Vietnam, in 1980, azolla was used in the northern and north central provinces on about 500 000 ha, about 14% of the irrigated rice growing area and 9% of the total rice area [50]. On the Red River Delta, azolla is grown for 40–60% of the irrigated spring rice [77].

In countries other than China and Vietnam, azolla use is incidental and limited. In the Philippines, farmers adopted azolla as green manure on more than 5 000 ha in South Cotabato in 1981 [68]. In 1983 it was used on 26 000 ha, although all the areas are not necessarily suitable for azolla growth. In Gambia, spontaneously growing azolla is incorporated as a green manure [108].

4 Limiting factors. A first requirement for azolla use is water availability and control. Azolla cannot withstand desiccation. Water should be in the field throughout cultivation. Additionally, because azolla grows from vegetative multiplication, inoculum must be maintained in nurseries all year and multiplied for distribution before field inoculation and multiplication. Those two requirements imply that an irrigation network and a network for inoculum conservation, production, and distribution are prerequisites for azolla utilization. This also imply that azolla adoption by farmers first depends on a governmental policy to establish such networks.

High temperatures retard azolla growth. Cool weather is a key to successful azolla utilization in Vietnam and China. However, successful growth of A. *pinnata* was observed at high temperatures in north Senegal [138]. Although azolla growth slows at temperatures higher than 30 °C, the major detrimental effect in relation to high temperature and humidity is the resulting high incidence of insect and fungus pests.

Among nutrients, P is most important. Success in spreading A. *pinnata* in South Cotabato, Philippines, was mainly due to a high level of available P in the soil. Watanabe and Espinas [157] reported that 25 ppm (Olsen P) in the soil is optimum for azolla and that P absorbing capacity of the soil should be less than 440 mg P/100 g soil. Such conditions are quite rare in rice soils and P fertilization for azolla growth usually is needed.

Economics of azolla use are very important. The technology used in Vietnam and China is labor intensive and could not be adopted in most rice growing countries. Kikuchi et al. [68] studied the economics of azolla use in the Philippines in South Cotabato, where azolla spreaded spontaneously and no P fertilizer and little labor were needed. They concluded that in that area economic return from azolla adoption, including cost savings in chemical fertilizers and weed control, was more than $35/ha at 1981 prices. Economic return from savings on inorganic fertilizer was about $10.

However, the authors clearly indicated that conditions in the studied area were exceptionally favorable and should be viewed realistically. They indicated that the economic potential of azolla is greatest where the opportunity cost of labor is low, and calculated that labor cost for azolla use becomes critical where agricultural wage rates approach $2/d. Insect control was also an important economic limitation. If more than 200 g carbofuran/ha (active ingredient) is needed to control insects, economic benefits are eliminated.

5 Conclusion. Azolla is N_2-fixing organic manure with similar N potential to that of legume green manures. It is easier to incorporate than other organic manures and grows well with rice in flooded conditions. Environmental and technological limitations are important but can be managed. Problems arising from inoculum conservation, multiplication, and transport could be solved to a large extent if azolla could be propagated from spores. Azolla sporulates but no method is known to induce sporulation. Possible use of A. *filiculoides* sporocarps for oversummering is being tested in China [72]. Temperature limitations and P requirements can be reduced by selecting cold or heat tolerant strains with low P requirements and by using P fertilizer. Labor costs do not apply in many rice growing countries. Among green manures, azolla is still less utilized than legumes but, contrary to legume use, azolla use is reported to be increasing, and many countries are evaluating it for popularization.

C Straw application

Straw incorporation is a traditional agricultural practice which has been primarily used to add nutrients and organic matter to the soil. A large volume of field data elucidates the effectiveness of organic amendments for increasing N fertility of rice soils. Beneficial effects on BNF were first demonstrated in small scale laboratory experiments [8, 100, 112, 162]. During the first *Nitrogen and rice* symposium at IRRI, Matsuguchi [81] presented data indicating that the root-free soil layer was the most important site for N fixation in a Japanese rice field. Additional research suggested that organic debris were an important site for heterotrophic BNF [145] and that straw application benefited photodependant BNF [83].

As a cultural practice, straw incorporation should be considered as an integrated management in which the beneficial effect on N fixation is only one of the components.

Rice straw contains about 0.6% N, 0.1% P, 0.1% S, 1.5% K, 5% Si, and 40% C. Because it is available on site in amounts varying from 2 to 10 t/ha, it is a convenient source of plant nutrients. Ponnamperuma [97] wrote that total nutrient content of straw in India, Philippines, and Sri Lanka in 1979 was more than twice that of chemical fertilizers used on rice in those countries. Straw also contains sugar, starches, celluloses, hemicelluloses, pectins, lignins, fats, and proteins (about 40%, as C, of the dry matter) that provide substrates for microbial metabolism (including N_2-fixing micro-organisms) in the soil.

Long-term experiments indicate that straw incorporation in lowland rice fields increases organic C and N, and available P, K, and Si. Straw is a slow release source of nutrients for rice. Undecomposed straw releases less N initially but more N later, than decomposed straw [20]. Peak absorption of N from incorporated straw was at middle plant growth stage. Plants recovered 25% N from straw after 130 d [20].

The yield advantage from straw incorporation rather than straw burning or removal is about 0.4 t/ha per season, and increases with time as soil fertility improves [97].

1 Potential for BNF

a Evidence of the process

In a 5 yr drum study with 3 soils, N increase from straw incorporation was computed to be 40 kg/ha per season, about 10 kg/ha per season more than the straw's N content. The extra N probably came from BNF by heterotrophs and phototrophs [97, 56].

b Effect on heterotrophic anaerobic BNF

In the early 1940s Jensen [62] wrote that when cellulosic material is added to soil 'a certain degree of anaerobiosis is necessary to bring about an active fixation (of N). This was confirmed by Mayfield and Aldworth [84]. Using 1 cm diameter, sandy clay loam soil aggregates amended with 2% wheat straw, they found that the only N_2-fixing bacteria that developed were anaerobes or facultative anaerobes, whether the aggregates were incubated under aerobic or anaerobic conditions, and that acetylene reducing activities (ARA) were similar under both conditions. Yoneyama et al. [162] reported that ARA of waterlogged soils was stimulated by incorporating rice straw at a relatively early stage of decomposition. Straw decreased the inorganic N and redox potential, making the environment favorable for anaerobic N_2-fixing bacteria. BNF was stimulated only when soils were waterlogged.

c Effect on other N_2-fixing organisms

The beneficial effect of straw applications is not limited to anaerobic heterotrophic fixers, but also applies to aerobic heterotrophs and phototrophs.

Straw application (5 and 10t/ha) was reported to enhance *Azotobacter* populations in submerged rice soil [101]. Magdoff and Bouldin [79] using a waterlogged sand matrix supplemented with a small amount of soil and mineral nutrients and enriched with cellulose, found fixation to be 10 to 15 times greater in the upper 2 to 3 mm of soil than in the lower layers. They suggested that products of anaerobic decomposition may be an energy source for aerobic BNF when brought under aerobic conditions by diffusion. Reddy and Patrick [105] observed a four fold increase of N_2-fixing activity in soils incubated under light after straw incorporation (0.4%). Similar results did not occur in darkness. However, the authors indicated that the increase in BNF in light incubated straw-enriched samples was unrelated to algal growth.

Matsuguchi and Yoo [83] measured ARA of straw fraction, root fraction and soil 7 w after transplanting in a soil with 8 t incorporated straw/ha. In the 0- 1 cm surface layer, photodependent and photoindependent ARA of straw were roughly 1000-fold and 100-fold higher than those in the soil fraction. In deeper layers the trend was similar. The same authors compared the effect on BNF of deep placement and topdressing of 8 t rice straw/ha in two soils. In all cases straw stimulated BNF. Stimulation was more marked for photodependent ARA in the 0−1 cm surface layer than for photoindependent ARA in the whole profile. Topdressing rice straw induced higher photodependent ARA and better rice growth than deep placement. Photodependent ARA in the topdressed straw was 10^3-fold more than in the soil fraction and was enhanced by N and herbicide application [83].

d Effect in the presence of N fertilizers

There are reports that straw incorporation may significantly reduce the inhibitory effect of N fertilizer on BNF.

Ammonium sulfate suppressed BNF with 5 t incorporated straw/ha but was not inhibitory with 10 t/ha [102]. Rhizosphere soil from plots receiving 3−6 t rice straw/ha exhibited more pronounced nitrogen fixing activity than the control. Forty to 80 kg mineral N/ha did not inhibit BNF with 6 t straw/ha [17]. Kalininskaya et al. [65] reported a maximum rate of BNF (ARA and ^{15}N) with N fertilizer and high straw application rates. They also reported that applying 100 to 200 kg N/ha did not inhibit BNF in the presence of straw [63].

On the other hand, Charyulu and Rao [19] reported that the inhibitory effect of ammonium sulfate increased with N concentration in small scale experiments with four soils enriched with 1% cellulose (Table 4). Similar results were presented by Rao [100].

e Quantification of fixed N

When analyzing experiments conducted to evaluate BNF in soils after straw (and other carbon sources) incorporation, it appears that all experiments are laboratory incubations of small samples of a few grams of soil (Table 4). In many experiments the quantity of straw incorporated was much higher than incorporation in the field. Assuming 300 t soil/ha, 3 t of straw incorporated per hectare equals 0.1% whereas 1–2% were commonly used in the experiments and sometimes as much as 20% (w/w) was incorporated [111].

Such experiments are hardly representative of the field conditions and seem to substantially overestimate amount of fixed N. For example, Charyulu et al. [18], using 5 g soil samples enriched with 0.5% rice straw, reported values ranging from 200 to 740 μg N fixed/g of soil in 100 d. Assuming 3000 t soil/ha of rice field, the above values correspond to 600 to 2220 kg N fixed/ha in 100 d. Under waterlogged conditions with 1 to 2% straw, 10 to 70 μg N/g soil were found to be fixed in 28 d [112], which equals 30–210 kg N fixed/ha in 28 d. High activity (45 mmol C_2H_2/g dry soil per h) were reported by Durbin and Watanabe [36] 4 d after incorporating 14 mg rice straw/g dry soil.

Results of larger scale expeiments show that a few mg N are fixed per gram of straw incorporated. Using ^{15}N, Rao [102] estimated that BNF by free-living bacteria in a flooded rice field was 7 kg N/ha in unamended soil and 25 kg N/ha in amended soil (5 t straw added; 3.6 mg N fixed per gram of straw). In saline takyr soils of Kazakh, BNF productivity was estimated as 5–10 mg N/g of applied straw [64]. Balance experiments by App and collaborators showed that about 5.5 mg N was fixed per gram of straw added [56].

When pooling data from Table 4 it appears that efficiency of heterotrophic BNF varies from 0.08 to 7.07 mg N fixed per gram of substrate (mainly straw) added. The average value of the 35 listed data is 2.39 mg N fixed per gram of substrate added, in about 1 mo.

2 Straw management in rice growing countries. Rice straw can be utilized or disposed by:
– direct return to the soil either as organic matter, with or without preliminary composting, or as ash after burning;

Table 4. Summary of the effects of carbon substrate incorporation on biological N_2 fixation

Soil	Substrate		Incubation		Method[b]	Efficiency: mg N fixed per g of substrate		Reference
	Nature	Concentration (%)	Conditions[a]	Duration (days)		Consumed	Added	
Dryland (5g)	Barley straw	5	Subm.	28	Kj.		1–2	8
		10	Subm.	28	Kj.		2.2–2.5	
		20	Subm.	28	Kj.		0.8–1.0	
		40	Subm.	28	Kj.		1–2	
Dryland (5g)	Straw	5	Subm.	30	Kj.		0.08	14
		16	Subm.	30	Kj.		2.15	
		32	Subm.	30	Kj.		1.26	
Soil dilutions nonamended previously amended	Glucose				Kj.		2–3	18
							3–6	
5 g soil	Cellulose straw	1	Subm.	20	^{15}N		1–2	19
							3–4	63
50 g of straw inoculated not inoculated	Wheat straw	100	Aero.	56		11.5	5.0	78
		100	Aero.	56		8.8	2.8	
	Glucose	1.5	Anaero.			12.3		91
	Glucose	2.5	Anaero.			14.2		
	Glucose	3.0	Anaero.			10.6		
	Mannitol	1.5	Anaero.			10.8		
	Mannitol	2.0	Anaero.			10.3		
	Mannitol	3.0	Anaero.			7.7		
5 g of soil with different pretreatments	Cellulose	1.0	Subm.	30	^{15}N		2–7	102

Table 4. Contd.

Soil	Substrate		Incubation		Method[b]	Efficiency: mg N fixed per g of substrate		Reference
	Nature	Concentration (%)	Conditions[a]	Duration (days)		Consumed	Added	
Meadow chernozem rice soil	Rice straw	1.0		30	^{15}N		0.53	100
		2.0		30	^{15}N		0.91	
	Cellulose	1.0	Subm.	30	^{15}N		1.72	103
		1.0	Anaero.		^{15}N	4.2–20		
Dark brown chernozenic soil (0.6 g)	Wheat straw	1	Subm.	28	^{15}N		6.7	112
		5	Subm.	28	^{15}N & Kj.		4.4	
2 g soil	Wheat straw	20	Subm.	28	Kj.		2.3	
	Wheat straw	20	Subm.	21	Kj.	16.1	2.2	111
2 g sand-clay + 0.1 g soil	Wheat straw	19	Subm.	21	Kj.	12.8	2.1	
2 g soil	Hemicellulose	5.0	Subm.	21	Kj.		2.4	
2 g soil	Straw residue	20	Subm.	21	Kj.		1.1	
2 g sand-clay + 10^8 cell C. pasteurianum	Glucose	0.5	Subm.	21	Kj.		1.5	
	Glucose	3.0	Subm.	21	Kj.		1.1	

[a]Subm.: submerged; Aero.: aerobiosis; Anaero.: anaerobiosis
[b]Kj.: Kjeldahl

- indirect return to the soil after such uses as bedding for cattle, mushroom culture, or mulch for a crop after rice;
- use as feed, fuel, or household roofing material; or
- industrial uses such as for paper, rope, bags, mats, or packing material.

The status of straw handling in rice growing countries was reviewed by Tanaka (131) who distinguished four major groups:
- countries where the major concern is to dispose of straw with
- minimum labor and where it is burned or incorporated – (USA, Southern European countries, Japan);
- countries where there is a belief that recycling organic matter makes soil more productive and where straw is harvested at the ground level and returned to the soil mainly as compost (China, Vietnam);
- countries where straw is mainly used as feed or fuel (India, Egypt); and
- countries where rice is frequently harvested at a rather high level and where much of the straw is returned to the soil, but although unintentionally (Malaysia, Indonesia, Philippines).

Tanaka's [131] survey indicated that burning is the most frequent straw removal-use practice. Araragi and Tangcham [3] reported that in Thailand, Burma, Philippines, Indonesia, and Malaysia, most rice straw is burned before starting the next cultivation. Straw incorporation is widely practiced only in China and before the summer rice crop in Vietnam. In some areas the straw is spread on the field and partially decomposed on moist soil or under water to facilitate incorporation. Heaping straw in mounds at threshing sites in the field is common in the Philippines and Indonesia. Straw decomposes slowly, largely aerobically, and is spread and easily incorporated at the beginning of the next season. This practice, however, reduces planted area and causes large nutrient losses by N denitrification, leaching, and C mineralization.

3 Limiting factors. Straw incorporation is an uncommon agricultural practice. Reasons are comprised of detrimental effects and socioeconomic factors.

a Detrimental effects

Adding straw to soils with high SO_4^{-2} and/or low in active Fe and Mn, such as acid sulfate soils and degraded saline soils may favor formation of H_2S, which is highly toxic to rice. Adding straw to neutral to alkaline soils, may markedly depress water soluble Zn, probably because of increased pCO_2, which leads to the formation of insoluble $Zn\ CO_3$. An increased HCO_3^- concentration in the soil solution also affects Zn root-to-shoot transport [162]. Yield of rice varieties susceptible to Zn deficiency has decreased after straw incorporation [58].

There are few reports on the effect of straw incorporation on BNF in relation to the release of toxic compounds such as organic acids and phenolic compounds. Phenolic compounds produced by decomposing rice straw inhibited the growth of and acetylene reduction by *Anabaena cylindrica* [109].. Such compounds also inhibited strains of *Rhizobium leguminosarum* and *R. japonicum* and reduced BNF by legumes *in situ*. This may explain the great reduction in soybean yields following rice that are observed in Taiwan when rice stubble is left in the field [110].

b Socioeconomic factors

Straw management is traditional, as are other agricultural operations. Where straw is utilized for definite purposes (feed, fuel, etc.) management changes cannot be expected without a demonstrated economic advantage. Straw is bulky and requires substantial labor to incorporate. Additionally, uniform incorporation with low power equipment is difficult. Where two rice crops are grown, straw disposal should be rapid and burning is the easiest method. Economics do not always favor incorporation of undecomposed straw. In a survey of straw management in the Philippines, Marciano et al. [80] observed that farmers do not follow practices that maximize the benefits of straw incorporation, despite the demonstrated beneficial effect on yield and soil fertility. Because incorporation of undecomposed straw is labor and power intensive in peak labor periods of land preparation and because the necessary time for decomposition before transplanting may delay crop establishment, Philippine farmers adopt management practices that require little labor, i.e., burn, stack/decompose, and stack/burn. Benefits of incorporating undecomposed straw were perceived by farmers to be uncertain and generally limited. Marciano et al. [80] calculated that 7–17% yield increase is necessary to make undecomposed straw incorporation more attractive to farmers than current practices. Incremental costs of incorporating undecompsed straw probably frequently exceeded the expected benefits, therefore farmers had little financial incentive for adoption.

4 Conclusion. Straw incorporation is an agronomic practice that returns significant quantities of nutrients to the soil and increases heterotrophic and autotrophic BNF. Data indicate that 2–3 kg N can be fixed per ton of straw incorporated. Limitations are primarily socioeconomic.

III Developing technologies

A Free living BGA

Among the N_2-fixing microorganisms, only BGA can generate photosynthate from CO_2 and water. This trophic independence makes BGA especially attractive as a biofertilizer. The agronomic potential of BGA was recognized in 1939 by De [28] who attributed the natural fertility of tropical rice fields to N_2-fixing BGA. Since then many trials have been conducted to increase rice yield by algal inoculation. Literature on BGA and rice was reviewed by

Roger and Kulasooriya [115] and BGA in tropical soils was reviewed by Roger and Reynaud [117].

1 Potential. Potential of BGA as a N source for rice can be described in three different ways:
- evaluation of BGA biomass and N content,
- measurement of N_2-fixing activity, and
- field experiments measuring BGA productivity and rice yield.

Records of BGA biomasses were summarized by Roger and Kulasooriya [115]. Fresh weight estimates range from a few kg to 16 t/ha and dry weight estimates from a few kg to 480 kg/ha. However, because of the variable dry matter, 0.5–5%; ash, 5–70%; and N, 2–13% contents of BGA field samples [58] such data are of little significance. Recent evaluations of artificially produced BGA blooms indicated standing biomass of N_2-fixing strains culminating at 150–250 kg dry weight/ha on an ash free basis, equalling 10–20 kg N/ha [59]. Those values may be considered to be the maximum standing biomass that can be expected in a rice field at blooming time. However, they underestimate the value for BNF, which is the result of the activity of a standing biomass and its turnover. No data are available on nutrient turnover rate from field-growing BGA.

Another way to roughly estimate BGA potential is to assume that all C input in the floodwater and surface soil is through BGA (which is an over-estimation). Saito and Watanabe [120], estimated an input of 0.6 t C in phytoplankton/crop per ha. Using that estimate and assuming a BGA C:N ratio ranging from 4 to 16 [58], the potential contribution of N_2-fixing BGA could be 37 to 150 kg N/ha per crop.

BNF by BGA has been most frequently studied by the acetylene reducing activity method, which may provide erroneous results [75]. ARA variations during the day and the growing cycle can be rapid and important, and moreover, ARA has a log-normal distribution [118]. Therefore, many replicates and very frequent measurements are needed to satisfactory measure total ARA.

However, this tedious work will provide an imprecise evaluation of the N_2-fixing activity (NFA) because the conversion factor of acetylene-N is not constant and must be determined [95]. Few reliable estimations of ARA have been published, and the number of measurements and replicates have been generally too low. Reported data on BNF related to BGA varied from a few to 80 kg N/ha and averaged 27 kg/ha per crop [115].

Field experiments on algal inoculation (algalization) provide indirect information on the overall potential of BGA. The most frequently used criterion for assessing the effects of algalization has been grain yield. Results of field experiments conducted mainly in India report an average 14% yield increase over the control corresponding to about 450 kg grain/ha per crop, where algal inoculation was effective. A similar increase was observed with and without N fertilizers. Because BNF is known to be inhibited by inorganic N the beneficial effect of algalization in the presence of N fertilizers was most frequently interpreted as resulting from growth-promoting substances produced by algae or by a temporary immobilization of added N, followed by a slow release through subsequent algal decomposition that permitted more efficient crop N utilization. Such interpretations have yet to be demonstrated.

Grain yield measurements suggest that algalization produces both a cumulative and residual effect. This was attributed to a build-up of organic N content and the number of BGA propagules in the soil, which facilitated the reestablishment of the BGA biomass. Several reports indicate an increase in organic matter and organic N. Algalization was also reported to increase aggregation status of the soil [125], waterholding capacity [127], and available P, total microflora, *Azotobacter, Clostridium,* and nitrifiers [47].

Thus, it appears that the potential of BGA has not been clearly quantified. Biomass measurements indicate that BGA have less potential than legumes and azolla. Comparison with N fertilizers indicated that algal inoculation may be equivalent to the application of 25–30 kg N/ha [142] but nothing is known about the relative importance of fixed N and other possible effects (auxinic effect, P solubilization, effects on soil properties and microflora, etc.) in the reported yield increase.

A classical statement in the reports on BGA inoculation is 'although the yields obtained in inoculated plots were higher, the difference between the yields of plots using and not using BGA was not significant'. This indicates that: the response to algal inoculation varies, the response is small, and the experimental error is larger than the response. The most common design for BGA inoculation experiments has been 4 × 4 m plots with 4 replicates which usually gives a coefficient of variation higher than 10% and a minimum detectable difference of 14.5% [44]. Such a value agrees with the average increase in yield reported after algal inoculation and confirms that a relatively low yield increase can be expected from this practice.

2 Methods of utilization. Research on methods for using BGA in rice cultivation, emphasizes algal inoculation (algalization) alone or together with agricultural practices favoring the growth of inoculated strains. This arose from the earlier belief that N_2-fixing strains were not normally present in many rice fields. It appears now that results concerning the occurrence of N_2-fixing BGA in rice fields are controversial. Watanabe and Yamamoto [146] found that only 5% of 911 soil samples from Asia and Africa harbored N_2-fixing species. Venkataraman [141] reported that 33% of 2213 soil samples from rice fields in India contained N_2-fixing strains. Okuda and Yamaguchi [90] reported the presence of N_2-fixing strains in 71% of the samples they collected in Japan. Reynaud and Roger [107] found N_2-fixing strains in 95% of the samples they collected in Senegal. In a survey of 40 rice fields in Thailand, Matsuguchi et al. [82] found BGA in all soils. In an ongoing survey of the Phillipine rice soils, we found N_2-fixing strains in all of the 79 samples collected (unpublished). N_2-fixing strains most probably are more common in rice fields than was previously thought. Unsuitable survey methodology, especially sampling method, probably caused the low values recorded [117]. Therefore, research should equally emphasize inoculation and indigenous strain enhancement.

a Algal inoculation

The methodology of BGA inoculum production was reviewed by Watanabe and Yamamoto [146] and Venkataraman [140]. Methods of field application were reviewed by Venkataraman [142]. Inoculum production in artificially controlled conditions was developed mainly in Japan where algalization is not used. Inoculum production under artificially controlled conditions is efficient but expensive. Open air soil culture, developed in India, is more simple, less expensive, and easily adoptable by the farmers. It is based on the use of a multistrain starter inoculum of *Aulosira, Tolypothrix, Scytonema, Nostoc, Anabaena,* and *Plectonema* provided by the 'All India Coordinated Project on Algae' [1]. The inoculum is multiplied by the farmer in shallow trays or tanks with 5–15 cm water, about 4 kg soil/m^2, 100 g triple superphosphate/m^2, and insecticide. If necessary, lime is added to correct the soil pH to about 7.0–7.5. In 1 to 3 w, a thick mat develops on the soil surface and sometimes floats. Watering is stopped and water in the trays is allowed to evaporate in the sun. Algal flakes are scraped off and stored in bags for use in fields. Using that method, the final proportion of individual strains in the algal flakes is unpredictable, but it is assumed that, because the inoculum is produced in soil and climatic conditions similar to those in the field, dominant strains will be the best adapted to the local conditions. The recorded rates of production of algal flakes in the open air soil culture range from 0.4 to 1.0 kg/m^2 in 15 d, indicating that in 2–3 mo a 2 m^2 tray can produce enough algal material to inoculate a 1 ha rice field. For transplanted rice, the algal inoculum is generally applied 1 w after transplanting. When rice is direct-seeded, seeds can be coated by mixing the algal suspension and 2–3 kg calcium carbonate per 10–20 kg seed and air-dried in the shade.

Recommendations for field application of dried algal inoculum (algal flakes) given by the All India Coordinated Project on Algae [1] indicate that:
- 8–10 kg of dry algal flakes applied 1 w after transplanting is sufficient to inoculate 1 ha, a larger inoculation will accelerate multiplication and establishment in the field;
- algalization can be used with high levels of commercial N fertilizer, but N application should be reduced by 30%;
- to benefit from the cumulative effect of algalization the algae should be applied for at least three consecutive seasons; and
- recommended pest-control measures and other management practices do not interfere with BGA establishment and activity in the fields.

b Methods to enhance indigenous BGA growth

The growth of N_2-fixing BGA in rice fields is most commonly limited by low pH, P deficiency, and grazer populations. Application of P and lime has frequently increased growth, particularly in acidic soils [153]. Increased BGA biomass has also been reported after insecticide application [46].

Recently, surface straw application was reported to benefit BGA growth and photodependent ARA [119]. This may be due to an increase of CO_2 in the photic zone, a decrease of mineral N and O_2 concentration in the flood-water, and the provision of microaerobic microsites by the straw. Increased CO_2 availability and low N concentration favor the growth of N_2-fixing BGA. Low O_2 concentration in the photic zone may increase specific N_2-fixing activity.

3 Current usage. Most of applied research on algal inoculation is conducted in India where a national program has been developed, the All-India Co-ordinated Project on Algae. To a lesser extent, applied research is also conducted in Burma, China, and Egypt.

Reports on the adoption of algal technology are controversial, but even considering the most optimistic evaluations, use of algal inoculation is restricted to very limited hectarage in a few Indian states and in Burma.

In a review on adoption of biofertilizers in India, Pillai [96] wrote: 'Apart from the work carried out at Research Stations very little organized work on development of the material for being adopted by the farmers has been taken up, especially in areas where it could be of potential benefit.'

In a review on biofertilizers, Subba Rao [129] wrote that the production capacity of BGA flakes in India was around 40 t/yr, which was approximately 0.01% of the total inoculum requirement for the country (40 t will inoculate 4 000 ha).

From the most recent extensive report on BGA field trials published by the Agricultural Economics Research Center of the University of Madras [135], it appears that despite an official radio and print publicity campaign, BGA use remains at the trial level and that in many cases inoculated algae did not multiply. Therefore, it seems appropriate to consider that this technology is more at an experimental level of large scale field testing than at a popularization stage.

4 Limiting factors. The major limiting factor to adopt algal inoculation is the lack of reliable technology for recommendation to farmers. Inoculum establishment is sporadic and the reasons for failure are frequently not known.

In reviewing BGA literature, it is surprising to observe the imbalance between the different topics. Taxonomy, morphology, micromorphology, physiology, and enzymology are highly documented and test tube BGA growth has been studied extensively. However, field studies are rare, most probably limited by lack of suitable methodology. Therefore, BGA ecology is still poorly understood.

The physiological characteristics of N_2 fixation desirable for strains suitable for field inoculation are known [128], but the selection of 'Super N_2-Fixing Strains' is meaningless unless they survive, develop, and fix N_2, as programmed, in rice fields.

As indicated by Gibson [41], virtually nothing is known of the attributes permitting introduced strains to colonize the various hostile environments to which they will be exposed. Similarly, the factors permitting the establishment of an N_2-fixing bloom of inoculated or indigenous strains still are unknown.

Low pH, low temperatures, and P deficiency limit BGA growth. However, because in some soils algalization is inefficient despite the addition of lime and phosphate [90], pH and available P are not the only limiting factors, but texture, organic matter content, CEC of saturated extracts, and total N are probably not important [130]. Grazing by invertebrate populations is an important biotic limiting factor [45]. Other possible limiting mechanisms such as antagonism, competition, etc. have been suggested, but their role is unclear. Low temperatures, heavy rains, and cloudy weather also have been reported to limit the inoculum establishment [115].

Inoculum quality also may be a limiting factor. In published methods of inoculum production, no tests of composition and viability have been included. We have shown that the density of colony-forming unit in BGA inocula may vary from 10^3 to 10^7/g of dry inoculum and that in some cases N_2-fixing strains are not dominant [59]. Some commercial inoculants also have been reported to have limited potential for BGA population enhancement [132]. Therefore, special attention must be paid to inocula quality.

Economics apparently do not limit BGA utilization. In a study of the economics of BGA use of 40 farmers in Tamil Nadu [135], no significant difference was found in the average per hectare cost of cultivation between crops using ($247) and not using ($246) BGA. The average return of BGA utilization was $4/ha.

When needed, grazer population can be controlled with inexpensive natural insecticides [46].

5 *Conclusion.* Blue-green algae seems to be a possible N source in neutral to alkaline soils with moderate to high P availability and low grazer incidence. BGA have less potential in terms of N_2-fixed than legumes or azolla, but need very limited additional inputs. Their usage is limited by technological problems governing inoculum quality and establishment and by generally low yield increases attributable to algal inoculation.

B Heterotrophic BNF associated with the rice plant

In 1929, Sen [124] reported the presence of N_2-fixing bacteria in rice roots, but his observation was overlooked. In 1961, Dobereiner and Ruschel [31] studied the growth stimulation of N_2-fixing bacteria in the rhizosphere of lowland rice. Their observations led to Dobereiner's idea that non-nodulated, non-leguminous plants can fix N through bacteria associated with roots [30]. The association between roots and bacteria was called rhizocoenosis.

In 1971, Rinaudo and Dommergues [113] and Yoshida and Ancajas [164] demonstrated N_2-fixing activity of wetland rice roots by using sensitive acetylene reduction assays. It was confirmed by $^{15}N_2$ incorporation [37, 60, 166] and N balance studies [2]. The basal portion of shoots also is a site of heterotrophic BNF [156]. Various groups of N_2-fixing bacteria were isolated from the same root samples using a non-selective medium [150] and a spermosphere model [5]. Ohta and Hattori [89] described an oligotrophic N_2-fixing bacteria that is abundant in rice roots. Bacteria genera isolated from rice roots are *Azospirillum* [70], *Beijerinkia* [31], *Pseudomonas* [7], *Enterobacteria* [71], *Klebsiella* [71], *Flavobacterium* [5], *Alcaligenes* [99], and *Agromonas* [89]. The N_2-fixing ability of the association has been proved, some major N_2-fixing organisms have been identified, and there have been some field and laboratory studies. However, there are many uncertainties about heterotrophic BNF in the rice rhizosphere. Major concerns of current research are: quantification of N_2 fixation, interactions between plant and bacteria, and varietal differences in the ability of rice to enhance associative BNF. Heterotrophic BNF in tropical soils was reviewed by Yoshida and Rinaudo [165].

1 Potential. In early studies, excised root assays were used to estimate potential N_2 fixation rate [144]. However, that technique was neither quantitative nor semiquantitative [137]. In situ or undistribed core assays are now used to measure ARA associated with plant [167].

In many assays ARA was highest at or near rice heading stage [6, 13, 121] and ranged from $0.3\,\mu$mol C_2H_4/plant per hour in temperate regions [121, 165] to $2\,\mu$mol C_2H_4/plant per hour in the tropics [6, 162]. Assuming (1) that ARA measured at heading stage continues for 50 d, (2) an acetylene/N conversion rate of 3:1, and (3) a plant density of $25/m^2$, the estimated N_2-fixing rate would be 0.8– 6 kg N/ha per cropping season. Extrapolation from ^{15}N incorporation experiments ranges from 1.3 to 7.2 kg N/ha per cropping season [37, 60, 166].

A rough estimation of the maximum value of heterotrophic N_2 fixation in the rhizosphere can be calculated using estimated C flow from the roots, but no data are available for rice. Sauerbeck and Johnen [123], growing wheat from the seedling to maturity under $^{14}CO_2$, estimated that C respired by microorganisms in the rhizosphere and converted to microbial biomass accounted for 4–5 times the remaining root C at harvest. Using this value and

and 0.2 t C/ha of roots at harvest [120], 1.0 t C/ha is estimated to pass through the microbial biomass in the rhizosphere. Assuming that all C is used for N_2 fixation (which does not happen) and 40 mg N as fixed/g C consumed, 40 kg N/ha would be the theoretical maximum of associative BNF.

Based on actual and potential estimates on N_2 fixation, it may be said that the potential of associative BNF is the least among N_2-fixing agents discussed in this paper.

2 Possible utilization methods. Manipulations of rice varieties and root-associated bacteria are possible methods to enhance heterotrophic BNF.

a Varietal differences

There are several reports on the varietal differences in the ability to support associative BNF (Nfs character). The differences were genetically analyzed by Iyama et al. [61]. Nothing is known, however, about the physiological basis of the apparent varietal differences. Dommergues [33] reported mutants with higher Nfs character than parents. In a pot experiment, 90 rice varieties grown in flooded conditions, including wild *Oryza* species, were screened for the ability to stimulate N gains by Kjeldahl assays [58]. The methodology used estimated gains due to heterotrophs and phototrophs. Varietal differences were found, and maximum N gains were 8 times the minimum gains. Correlation coefficients of N gains were high with total N uptake, total dry matter production, and daily dry matter production. This indicated that vigorously growing rice plants stimulated more BNF (character NFs).

b Microflora manipulation

Inoculation of N_2-fixing bacteria has been reported to increase growth and yield of rice [104]. As with BGA, interpreting the results is difficult in the absence of experimental data on inoculum establishment and its N_2-fixing activity. Here again the relative importance of N_2 fixed and other possible inoculation effects must be evaluated. Results obtained by O'Hara et al. [88] who inoculated Nif^+ and Nif^- strains of *Azospirillum* on maize, indicated that observed growth stimulation was not because of N_2 fixation. Watanabe and Lin [158], using the ^{15}N dilution method, observed no difference in ^{15}N abundance between rice plants inoculated with *Azospirillum* or *Pseudomonas* sp. and non inoculated plants despite growth stimulation by inoculation.

3 Prospect. The idea of breeding varieties higher in Nfs is attractive because it would enhance BNF without additional cultural practices. However, a prerequisite to Nfs breeding is the availability of a rapid screening technique. Acetylene reducing activity was used for most of varietal screenings, however, it is time consuming because it needs several measurements during the plant growth cycle. Additionally, the high variability of the measurements may mask varietal differences. ^{15}N dilution may be used for screening and genetic studies, but for that purpose, reference varieties with low N_2 fixation stimulation ability must be identified.

N balance studies have shown that vigorously growing varieties tend to have high associative N_2 fixation ability. If this trait, a characteristic of traditional tall indica rices, is essential, the dilemma comes in choosing between the high yielding character associated with short satured plant types and the N_2 fixation stimulating trait associated with tall stature. However, IR42, a high yielding variety, high in stimulating N_2-fixation gives hope of possible selection for high yielding, high N_2-fixing varieties [58].

Another approach is microflora manipulation. If the positive effect of inoculation on plant growth is at least partly because of enhanced BNF, the selection or the improvement of N_2-fixing strains is the first possibility. N_2-fixing ability of bacteria isolated from rice may be enhanced by recombinant DNA techniques. The idea to develop derepressed N_2-fixing bacteria in which BNF is not suppressed by NH_4^+ appears to be attractive. However, it is not known if such derepressed bacteria can compete with non N_2-fixing ammonium utilizing microflora. A related problem is how to select competitive strains with the ability to live in the rhizosphere of inoculated plants. A better knowledge of root bacteria relationships is a prerequisite to this work.

No cultural practice are known to enhance the associative BNF process. Associative BNF seems to be lower in acidic soils [22, 27] and to be less sensitive to N fertilizer application than other N_2-fixing systems [155].

In the next few years, associative BNF may be understood by using a simple system like monoaxenic culture. Enormous effort will be needed to develop intentional utilization and enhancement of the process. As Postgate [98] stated, 'yet it would be as foolish to abandon the project of establishing productive rhizocoenoses with cereals as it was to expect such an association to mature in just a couple of years of research.'

IV General conclusion

Spontaneous BNF has permitted moderate but constant rice yield, around 2 t/ha, without N fertilization. Through traditional management practices such as green manuring with legumes or azolla, BNF contribution to soil N fertility has been substantially enhanced, thus producing higher yields (2–4 g/ha). Although all groups of N_2-fixing microorganisms have been shown to inhabit rice fields, and their possible agronomic use has been demonstrated, technologies available for popularization are still limited to practices based on the production of a large biomass followed by its incorporation, namely legumes, azolla, and straw incorporation.

Legumes and azolla have best known potential as N_2-fixers in rice fields, with recorded 50–100 kg N fixed/ha per crop. Current usage of separate-crop legumes is decreasing. Use of azolla, grown with rice, is increasing. Legume use is limited by socioeconomic factors whereas azolla technology is limited by socioeconomic and environmental factors. Straw incorporation, besides providing soil nutrients, increases heterotrophic BNF by 2–4 kg N/t of straw incorporated. Its use is limited by socioeconomic factors.

Other N_2-fixing systems are unused or little-used by farmers. Free living BGA have less potential than legumes and azolla (around 30 kg N/ha per crop) but their use is promising because little additional labor is required. However, algal inoculation is still at a research level in most of the rice growing countries. Factors involved in yield increase reported after algal inoculation, factors leading to the establishment of a bloom, and the general ecology of BGA in rice fields are still poorly understood. Therefore, algal inoculation cannot be confidently recommended yet.

There are evidences that some rice varieties promote heterotrophic BNF. However, no technology has yet been developed to utilize heterotrophic N_2 fixation associated with rice. The overall impression from experiments design to enhance the process by inoculation is not one of a great success.

Biological N_2 fixation has potential where N fertilizer is unavailable. Data on hectarage of rice grown without fertilizer are not available, but assuming that there is a correlation between water management and fertilizer use and considering that only about 30% of riceland in Southeast Asia is irrigated, it appears that the hectarage of nonchemically fertilized rice may represent a very large area.

A common characteristic of the BNF technologies currently adopted by farmers is intensive labor use. They are most often used under socio-economic conditions where labor intensive practices are economically feasible or where economics is not a major factor. In the future it is unlikely that BNF could be an exclusive N source for producing high yields under economically feasible conditions.

However, this does not mean that BNF technology has potential only where N fertilizer is unavailable or unaffordable, and that BNF will only produce low to moderate yields. Most probably the future of BNF in rice cultivation is in integrated management. A better knowledge of the microbiology and the ecology of rice fields will encourage high rice yields through a more efficient usage of chemical fertilizers and the simultaneous utilization of BNF.

When considering that:
- the efficiency of nitrogenous fertilizers is 30–50% or less [24];
- a large part of N not recovered in rice is lost; and that
- N fertilizer is applied so that it inhibits major components of the N_2-fixing biomass,

it appears that it is not a fine tuning of N fertilizer management in lowland rice that is needed but drastic changes in present fertilization concepts. Urea

70

deep placement [29], which significantly decreases losses of N by volatilization and does not inhibit photodependent BNF [116] is a good example of the kind of technology that must be developed for integrated management of BNF and chemical fertilizers.

When considering current usage by farmers, it appears that BNF is purposefully used in only a small percentage of rice fields in a few countries and that rice farmers are far from realizing its potential. This underutilization is due to ecological and socioeconomic factors and lack of technology development and knowledge. On a short-to medium-term basis, BNF has underexploited potential where N fertilizers are not available or affordable. On a long term basis, BNF integrated management should permit high yields with lower N fertilizer application.

References

1. All India coordinated project on algae (1979) Algal biofertilizer for rice. India Agric Res Inst. New Delhi, 61 pp
2. App A, Watanabe I, Alexander M, Ventura WV, Daez C, Santiago T and De Datta SK (1980) Nonsymbiotic nitrogen fixation associated with the rice plant in flooded soils. Soil Sci 130:283–289
3. Araragi M and Tangcham B (1979) Effect of rice straw on the composition of volatile soil gas microflora in the tropical paddy field. Soil Sci Plant Nutr 25(3): 283–295
4. Aston PJ (1974) Effect of some environmental factors on the growth of *Azolla filiculoides* Lam. pp 124–138. In Orange River Progress Report. Institute for Environmental Sciences, Univ. O.F.S., Bloemfontein. South Africa
5. Bally R, Thomas-Bauzon D, Heulin T and Balandreau J (1983) Determination of the most frequent N$_2$-fixing bacteria in a rice rhizosphere. Can J Microbiol 29: 881–887
6. Barraquio WL, De Guzman MR, Barrion M and Watanabe I (1982) Population of aerobic heterotrophic nitrogen-fixing bacteria associated with wetland and dryland rice. Applied Environ. Microbiol 43(1):124–128
7. Barraquio WL, Ladha JK and Watanabe I (1983) Isolation and denitrification of N$_2$-fixing *Pseudomonas* associated with wetland rice. Can J Microbiol 29:867–873
8. Barrow NJ and Jenkinson DS (1962) The effect of waterlogging on fixation of nitrogen by soil incubated with straw. Plant and Soil 16(2):258–262
9. Becking JH (1979) Environmental requirements of Azolla for use in tropical rice production, pp. 345–373. In Nitrogen and Rice, The International Rice Research Institute, Los Banos, Philippines
10. Beri V and Meelu O (1981) Substitution of N through green manure in rice. Indian Farming. May 1981
11. Bhardwaj SP, Prasad SN and Singh G (1981) Economizing N by green manures in rice wheat rotation. Ind J Agric Sci 51:86–90
12. Biswas TD, Roy MR and Sahu BN (1970) Effect of different sources of organic manures on the physical properties of the soil growing rice. J Ind Soc Soil Sci 18: 233–242
13. Boddey RM and Ahmed N (1981) Seasonal variations in nitrogenase activity of various rice varieties measured with *in situ* acetylene reduction technique in the field. In Associative N$_2$-fixation, Vol. 2:220–229, Vose, PB and Ruschel AP eds, CRC press
14. Bremner JM and Shaw K (1958) Denitrification in soil. Part I. Methods of investigation, J Agr Sci 52:40–52
15. Buresh RJ, Casselman ME and Patrick WH Jr (1980) Nitrogen fixation in flooded soil systems, a review. Adv Agron 23:149–192

16. Chari RV (1957) Agricultural development in Madras State. World Crops 9:33–37
17. Charyulu PBBN, Nayak BN and Rao VR (1981) ^{15}N incorporation by rhizosphere soil: influence of rice (*O. sativa*) variety, organic matter and combined nitrogen. Plant and Soil 59(3):399–406
18. Charyulu PBBN, Ramakrishna C and Rao VR (1978) Facultative symbiotic nitrogen-fixing associations in rice soil in India. Proc Indian Acad Sci 87 B(10): 243–246
19. Charyulu PBBN and Rao VR (1979) Nitrogen fixation in some Indian rice soils. Soil Sci 128(2):86–89
20. Chatterjee BN, Singh RJ, Pal A and Maiti (1979) Organic manures as substitutes for chemical fertilizer on high yielding rice varieties. Indian J Agric Sci 49(3):188–192
21. Chen S (1980) Green manures in multiple cropping systems in China. Soils and Fertilizer Institute, Liaoning, Academy of Agricultural Sciences, China, Quoted by Pandey and Morris, 1983
22. Cholitkul W, Tangcham B, Sangtong P and Watanabe I (1980) Effect of phosphorus on N_2-fixation measured by field acetylene reduction technique in Thailand long term fertility plots. Soil Sci Plant Nutr 26:291–299
23. Cowardin LM, Carter V, Golet FC and LaRoe ET (1979) Classification of wetland and deepwater habitats of the United States. Fish and Wildlife Service, U.S. Gov't. Printing Office, Washington, D.C. 103 pp
24. Craswell ET and De Datta SK (1980) Recent development in research nitrogen fertilizers for rice. IRPS no 49. The International Rice Research Institute, Los Banos, Philippines
25. Dao The Tuan and Tran Quang Thuyet (1979) Use of Azolla in rice production in Vietnam, pp. 395–405. In Nitrogen and Rice, The International Rice research Institute, Los Banos, Philippines
26. Dargan KS and Chillar RK (1975) Central Soil Salinity Research Institute, Karnal, India Annual Report 53–54, quoted by Venkataraman 1984
27. Daroy ML and Watanabe I (1982) Nitrogen fixation by wetland rice grown in acid soil, Kalikasan Philipp J Biol 11(2–3):339–348
28. De PK (1939) The role of blue-green algae in nitrogen fixation in rice fields. Proc R Soc Lond 127 B, 121–139
29. De Datta SK, Fillery IRP and Craswell ET (1983) Results from recent studies on nitrogen fertilizer efficiency in wetland rice. Outlook in agriculture 12(3):125–134
30. Dobereiner J and Day JM (1975) Nitrogen fixation in the rhizosphere of tropical grasses. pp 39–56. Stewart WDP ed In Nitrogen fixation by free-living microorganisms, Cambridge Univ Press
31. Dobereiner J and Ruschel AP (1962) Inoculation of rice with N_2-fixing genus Beijerinckia Derx. Rev Brasil Biol 21:397–407
32. Dommergues Y (1978) Microbial activity in different types of microenvironments in paddy soils, pp 451–466. In Environmental Biogeochemistry and Geomicrobiology, Vol 2, Krumsbein WE ed Ann Arbor Pub Michigan
33. Dommergues Y (1978) Impact on soil management and plant growth. In Interactions between non-pathogenic soil microorganisms and plants, pp 443–458, Dommergues Y and Krupa S, eds Elsevier
34. Dreyfus BL and Dommergues YR (1980) Non inhibition de la fixation d'azote atmospherique chez une legumineuse a nodules caulinaires, *Sesbania rostrata*. CR Acad Sci Paris, D 291:767–770
35. Dreyfus BL and Dommergues YR (1981) Nitrogen-fixing nodules induced by Rhizobium on the stem of the tropical legume *Sesbania rostrata*. FEMS Microb Letters 10:313–317
36. Durbin KJ and Watanabe I (1980) Sulfate-reducing bacteria and nitrogen fixation in flooded rice soils. Soil Biol Biochem 12:11–14
37. Eskew DL, Eaglesham ARJ and App AA (1981) Heterotrophic N_2-fixation and distribution of newly fixed nitrogen in a rice-flooded soil system. Plant Physiol 68: 48–52
38. FAO (1978) China: azolla propagation and small-scale biogas technology, FAO Soils Bulletin No 41

72

39. Fried M, Danso SKA and Zapata F (1983) The methodology of measurement of N_2 fixation by nonlegumes as inferred from field experiments with legumes. Can J Microbiol 29:1053–1062
40. Ghose RLM, Ghatge MB and Subramanyan V (1956) Rice in India – Indian Council of Agricultural Research, New Delhi
41. Gibson AH (1981) Some required inputs from basic studies to applied nitrogen fixation research, Pages 6–7. In Current perspective in nitrogen fixation. Gibson A and Newton WE eds, Australian Academy of Science
42. Gibson AH, Dreyfus BL and Dommergues YR (1982) Nitrogen fixation by legumes in the tropics, pp. 37–73. In Microbiology of tropical soils and plant productivity, YR Dommergues and HG Diem eds, M Nijhoff and W Junk
43. Gomez AA and Zandstra HG (1977) An analysis of the role of legumes in multiple cropping systems. In Exploiting the legume-rhizobium symbiosis in tropical agriculture. Vincent JM, Whitney AS and Bose J eds, University of Hawaii, USAID
44. Gomez KA (1972) Techniques for field experiment with rice. International Rice Research Institute, Los Banos, Philippines
45. Grant I and Alexander M (1981) Grazing of blue-green algae (Cyanobacteria) in flooded soils by Cypris (Ostracoda), Soil Sci Soc Amer J 45:773–777
46. Grant IF, Tirol AC, Aziz T and Watanabe I (1983) Regulation of invertebrate grazers as a means to enhance biomass and nitrogen fixation of Cyanophyceae in wetland rice fields. Soil Sci Soc Amer J 47:669–675
47. Ibrahim AN, Kamel M and El-Sherbeny M (1971) Effect of inoculation with alga *Tolypothrix tenuis* on the yield of rice and soil nitrogen balance. Agrokem Talagtan 20:389–400
48. ICAR (1977) Indian Council of Agricultural Research. New Delhi, India. Annual Report for 1977
49. International Institute of Agriculture (1936) Use of leguminous plant in tropical countries as green manure, as cover and as shade, pp 64–68
50. International Rice Research Institute (1982) Report on the INSFFER Azolla study tour in Vietnam, 20 Jan–Feb 1982, IRRI, Los Banos, 66 pp
51. International Rice Research Institute (1963) Annual Report for 1962
52. International Rice Research Institute (1966) Annual Report for 1965
53. International Rice Research Institute (1980) Reports on the 1st trial of Azolla in rice INSFFER 1979, Los Banos, Laguna, Philippines
54. International Rice Research Institute (1981) Reports on the 2nd trial of Azolla in rice INSFFER 1980, Los Banos, Laguna, Phillipines
55. International Rice Research Institute (1982) Reports on the 3rd trial of Azolla in rice INSFFER 1981, Los Banos, Laguna, Philippines
56. International Rice Research Institute (1982) Annual Report for 1981
57. International Rice Research Institute (1983) Reports on the 4th trial of Azolla in rice INSFFER 1982, Los Banos, Laguna, Philippines
58. International Rice Research Institute (1983) Annual Report for 1982
59. International Rice Research Institute (1984) Annual Report for 1983
60. Ito O and Watanabe I (1980) Fixation of dinitrogen-15 associated with rice plant. Appl Environ Microbiol 39:554–558
61. Iyama S, Sano Y and Fujii T (1983) Diallel analysis of nitrogen fixation in the rhizosphere of rice. Plant Sci Letters 30:129–135
62. Jensen HL (1941) Nitrogen fixation and cellulose decomposition by soil microorganisms. III. Proc Linnean Soc N S Wales 66:239–249
63. Kalininskaya TA, Miller UM, Belov UM and Rao VR (1977) [15]Nitrogen studies of the activity of non symbiotic nitrogen fixation in rice field soils of Krasnodar Territory. Izv Akad Nauk USSR Ser Biol 4:565–570 (in Russian)
64. Kalininskaya TA, Petrova AN, Nelidov SN, Miller YM and Belov YM (1980) Nitrogen fixation in Kazahk SSR, USSR, saline takyr soils under rice cultivation. Izv Akad Nauk SSR Ser Biol 0(5):747–753 (in Russian)
65. Kalininskaya TA, Rao VR, Volkova TN and Ippolitov LT (1973) Nitrogen-fixing activity of soil under rice crop studied by acetylene reduction assay. Microbiol 42: 426–429

66. Kanazawa N (1984) Trends and economic factors affecting organic manures in Japan. pp 557–567 In Organic Matter and Rice, The International Rice Research Institute, Los Banos, Philippines

67. Khan MM (1983) A primer on Azolla production and utilization in agriculture – UPLB – PCARRD – SEARCA, Los Banos, Philippines 143 pp

68. Kikuchi M, Watanabe I and Haws LD (1984) Economic evaluation of Azolla use in rice production. pp 569–592 In Organic Matter and Rice, The International Rice Research Institute, Los Banos, Philippines

69. Koyama T and App AA (1979) Nitrogen balance in flooded rice soils. pp 95–104 In Nitrogen and Rice, The International Rice Research Institute, Los Banos, Philippines

70. Ladha JK, Barraquio WL and Watanabe I (1982) Immunological techniques to identify Azospirillum associated with wetland rice. Can J Microbiol 28(5):478–485

71. Ladha JK, Barraquio WL and Watanabe I (1983) Isolation and identification of nitrogen-fixing *Enterobacter cloacae* and *Klebsiella planticola* associated with rice plants. Can J Microbiol 29:1031–1038

72. Li Shi Ye (1984) Azolla in the paddy fields of eastern China, pp 169–178, In Organic Matter and Rice, The International Rice Research Institute, Los Banos, Philippines

73. Li ZX, Zu SX, Mao MF and Lumpkin TA (1982) Study on the utilization of eight Azolla species in agriculture (in Chinese), Zhangguo Nongye Kexue 1:19–27

74. Liu Chung-chu (1979) Use of Azolla in rice production in China, pp. 375–394, In Nitrogen and Rice, The International Rice Research Institute, Los Banos, Philippines

75. Lowendorf HS (1982) Biological nitrogen fixation in flooded rice fields. Cornell Int Agric Mimeogr 96, Nov 1982, 75 pp

76. Lumpkin TA and Plucknett DL (1980) Azolla: botany, physiology and use as a green manure. Economic Bot 35(2):111–153

77. Lumpkin TA and Plucknett DL (1982) Azolla as a green manure, Westview Tropical Agriculture Series, Westview Press, Boulder Co, USA 230 pp

78. Lynch JM and Harper HT (1983) Straw as a substrate for cooperative nitrogen fixation. J Gen Microbiol 129:251–253

79. Magdoff FR and Bouldin DR (1970) Nitrogen fixation in submerged soil-sand-energy material media and the aerobic-anaerobic interface. Plant and Soil 33:49–61

80. Marciano UP, Mandac AM and Flinn JC (1983) Rice straw management in the Philippines. Paper presented at the 14th Annual Scientific Meeting of the Crop Science Society of the Philippines, 2–4 May 1983. IRRI, Los Banos, Philippines

81. Matsuguchi T (1979) Factors affecting heterotrophic nitrogen fixation in submerged soils, pp 207–222, In Nitrogen and Rice, The International Rice Research Institute, Los Banos, Philippines

82. Matsuguchi T, Tangcham B and Patiyuth S (1974) Free-living nitrogen fixers and acetylene reduction in tropical rice fields, Jpn Agric Res Q 8(4), 253–256

83. Matsuguchi T and Yoo ID (1981) Stimulation of phototrophic N_2 fixation in paddy fields through rice straw application, pp 18–25, In Nitrogen cycling in South-East Asian Wet Monsoonal Ecosystems, Wetselaar R, Simpson JR and Rosswall T eds, Canberra, Austral Acad Sci

84. Mayfield CI and Aldworth RL (1974) Acetylene reduction in artificial soil aggregates amended with cellulose, wheat straw and xylan. Can J Microbiol 20:1503–1507

85. Meelu OP and Rekki RS (1981) Mung straw management and nitrogen economy in rice culture. Int Rice Res Newsl 6 (4)21

86. Moore AW (1969) Azolla: biology and agronomic significance. Bot Rev 35:17–35

87. Morris RA and Pandey RK (1983) Integrated nitrogen management. The International Rice Research Institute, Internal Program Review Jan 21, 7 pp

88. O'Hara GW, Davey MR and Lucas JA (1981) Effect of inoculation of *Zea mays* with *Azospirillum brasilense* strains under temperate conditions. Can J Microbiol 27:871–877

89. Ohta H and Hattori T (1983) *Agromonas oligotrophica* gen. nov., a nitrogen fixing oligotrophic bacterium. Antonie van Leeuwenhoek 49:429–446
90. Okuda A and Yamaguchi M (1952) Algae and atmospheric nitrogen fixation in paddy soils. II. Relation between the growth of blue-green algae and physical or chemical properties of soil and effect of soil treatments and inoculation on the nitrogen fixation. Mem Res Inst Food Sci 4:1–11
91. O'Toole P and Knowles P (1973) Efficiency of acetylene reduction (nitrogen fixation) in soil: effect of type and concentration of available carbohydrate. Soil Biol Biochem 5:789–797
92. Palacpac AC (1982) World rice statistics. The International Rice Research Institute, Los Banos, Philippines 152 pp
93. Pandey RK and Morris RA (1983) Effect of leguminous green manuring on crop yields in rice-based cropping systems. Paper presented at the International Rice Research Conf. April 23–28, 1983, International Rice Research Institute, Los Banos, Philippines
94. Patnaik S and Rao MV (1979) Sources of nitrogen for rice production. pp 25–44. In Nitrogen and Rice, The International Rice Research Institute, Los Banos, Philippines
95. Peterson, RB and Burris RH (1976) Conversion of acetylene reduction rates to nitrogen fixation rates in natural populations of blue-green algae. Anal Biochem 73:404–410
96. Pillai KG (1980) Biofertilizers in rice culture. Problems and prospects for large scale adoption. AICRIP publ no 196
97. Ponnamperuma FN (1984) Straw as a source of nutrients for wetland rice. pp 117–136 In Organic Matter and Rice, The International Rice Research Institute, Los Banos, Philippines
98. Postgate J (1980) Nitrogen fixation in perspective. pp 423–435 In Nitrogen fixation, Stewart WDP and Gallon JR eds, Academic Press
99. Qui YS, Zhou SP and Mo XZ (1981) Study of nitrogen fixing bacteria associated with rice root. I. Isolation and identification of organisms. Acta Microbiologica Sinica 21:468–472
100. Rao VR (1976) Nitrogen fixation as influenced by moisture content, ammonium sulphate and organic sources in a paddy soil. Soil Biol Biochem 8:415–448
101. Rao VR (1977) Effect of organic and mineral fertilizers on *Azotobacter* in flooded rice fields. Curr Sci 46(4):118–119
102. Rao VR (1980) Changes in nitrogen fixation in flooded paddy field soil amended with rice straw and ammonium sulfate. Riso 28(1):29–34
103. Rao VR, Kalininskaya TA and Miller UM (1973) The activity of non symbiotic nitrogen fixation in soils of rice field studied with ^{15}N. Mikrobiologiya 42:729–734
104. Rao VR, Nayak DN, Charyulu PBD and Adhya TK (1983) Yield response of rice to root inoculation with *Azospirillum*. J Agric Sci UK 100(3):689–691
105. Reddy KR and Patrick WH Jr (1979) Nitrogen fixation in flooded soil. Soil Sci 128(2):80–85
106. Rerkasem K and Rerkasem B (1984) Organic manures in intensive cropping system. pp 517–531. In Organic Matter and Rice, The International Rice Research, Institute, Los Banos, Philippines
107. Reynaud PA and Roger PA (1978) N_2-fixing algal biomass in Senegal rice fields. Ecol Bull, Stockholm 26:148–157
108. Reynaud PA (1982) The use of *Azolla* in West Africa. pp 565–566 In Biological Nitrogen Fixation Technology for Tropical Agriculture, Graham PH and Harris SC eds, Cali, Colombia
109. Rice EL, Chi-Yung Lin and Chi-Yung Huang (1980) Effect of decaying rice straw on growth and nitrogen fixation of a blue-green alga (*Anabaena cylindrica*). Bot Bull Acad Sin 21(2):111–118 Taipei
110. Rice EL, Chu-Yung Lin and Chi-Ying Huang (1981) Effect of decomposing rice straw on growth and nitrogen fixation by Rhizobium. J Chem Ecol 7(2):333–344
111. Rice WA and Paul EA (1972) The organisms and biological processes involved in asymbiotic nitrogen fixation in waterlogged soil amended with straw. Can J Microbiol 18:715–723

75

112. Rice WA, Paul EA and Wetter LR (1967) The role of anaerobiosis in asymbiotic nitrogen fixation. Can J Microbiol 13:829–836
113. Rinaudo G and Dommergues Y (1971) The validity of acetylene reduction method for the determination of nitrogen fixation in rice rhizosphere. Ann Inst Pasteur 121:93–99
114. Rinaudo G, Dreyfus B and Dommergues YR (1981) *Sesbania rostrata* green manure and the nitrogen content of rice crop and soil. Soil Biol Biochem 15:111–113
115. Roger PA and Kulasooriya SA (1980) Blue-green algae and rice. The International Rice Research Institute, Los Banos, Philippines 112 pp
116. Roger PA, Kulasooriya SA and Craswell ET (1980) Deep placement: a method of nitrogen fertilizer application compatible with algal nitrogen fixation in wetland rice soils. Plant and Soil 57:137–142
117. Roger PA and Reynaud PA (1982) Free living blue-green algae in tropical soils. pp 147–168. In Microbiology of tropical soils and plant productivity, Dommergues Y and Diem H eds, Martinus Nijhoff La Hague
118. Roger PA, Reynaud PA, Rinaudo GE, Ducerf PE and Traore TM (1977) Log-normal distribution of acetylene-reducing activity *in situ*. Cah ORSTOM Ser Biol 12:133–140 (in French, English summary)
119. Roger PA, Tirol A, Grant I and Watanabe I (1982) Effect of surface application of straw on phototrophic nitrogen fixation. Int Rice Res Newsl 7(3):16–17
120. Saito M and Watanabe I (1978) Organic matter production in rice field flood water. Soil Sci Plant Nutr 24(3):427–440
121. Sano Y, Fujii T, Iyama S, Hirota Y and Komagata K (1981) Nitrogen fixation in the rhizosphere of cultivated and wild rice strains. Crop Sci 21:758–761
122. Sanyasiraju M (1952) Role of organic manures and inorganic fertilizers in soil fertility. Madras Agric J 39:132–139
123. Sauerbeck D and Johnen B (1976) Der Umsatz von Pflanzenwurzeln im Laufe der Vegetationsperiode und dessen Beitrag zur Bodenatmung. Z. Pflanzenern Bodenk 90:315–328
124. Sen MA (1929) Is bacteria associations a factor in nitrogen assimilation by rice plant? Agric J India 24:229–232
125. Shield LM and Durrel LW (1964) Algae in relation to soil fertility. Bot Rev 30:93–128
126. Singh NT (1984) Green manures as source of nutrients in rice production. pp 217–228 In Organic Matter and Rice, The International Rice Research Institute, Los Banos, Philippines
127. Singh RN (1961) Role of blue-green algae in nitrogen economy of Indian agriculture. Indian Council of Agric Res New Delhi, 175 pp
128. Stewart WDP, Rowel P, Ladha JK and Sampaio MJA (1979) Blue-green algae (Cyanobacteria) – some aspects related to their role as sources of fixed nitrogen in paddy soils. pp 263–285 In Nitrogen and Rice, The International Rice Research Institute, Los Banos, Philippines
129. Subba Rao NS (1982) Biofertilizer – interdisciplinary science reviews. 7(3):220–229
130. Subramanyan R, Manna GB and Patnaik S (1965) Preliminary observations on the interaction of different rice soil types to inoculation of blue-green algae in relation to rice culture. Proc Indian Acad Sci Sec B 62(4):171–175
131. Tanaka A (1973) Methods of handling the rice straw in various countries. Int Rice Comm Newsl 22(2):1–20
132. Tiedman AR, Lopushinsky W and Larsen HJ (1980) Plant and soil responses to a commercial blue-green algae inoculant. Soil Biol Biochem 12:471–475
133. Tiwari KN, Pathak AN and Hari Ram (1980) Green manuring in combination with fertilizer nitrogen on rice under double cropping system in an alluvial soil. J Ind Soc Soil Sci 28:162–169
134. Trebuil G (1983) Le systeme de mise en valeur agricole du milieu et son evolution recente dans la region de Sathing Phra – Sud Thailande Juillet 1983 – 101 pp + annexes. Universite Prince de Songala – Project de recherches sur les systemes de production agricole. Publ No 2

135. University of Madras (1982) Blue-green algae as a source of bio-fertilizers in Thanjauur district – Tamil Nadu. Res Study No 74 of the Agric Econ Res Center, Multigr 79 pp
136. Vachhani MB and Murthy KS (1964) Green manuring for rice. ICAR Res Report Ser No 17 Ind Council Agric Res New Delhi 50 pp
137. Van Berkum and Bohlool BB (1980) Evaluation of nitrogen fixation by bacteria in association with roots of tropical grasses. Microbiol Rev 44:491–517
138. Van Hove C, Diara HF and Godard P (1982) Azolla in West Africa, Azolla project. WARDA
139. Venkataraman A (1984) Development of organic matter-based agricultural systems in South Asia. pp 57–70 In Organic Matter and Rice, International Rice Research Institute, Los Banos, Philippines
140. Venkataraman GS (1972) Algal biofertilizers and rice cultivation. Today and Tomorrow's Printers, Faridabad (Haryana) 75 pp
141. Venkataraman GS (1975) The role of blue-green algae in tropical rice cultivation. pp 207–218 In Nitrogen fixation by free-living microorganisms, Stewart WDP ed, Cambridge Univ Press
142. Venkataraman GS (1981) Blue-green algae for rice production – A manual for its promotion. FAO Soils Bull No 46 102 pp
143. Ventura W and Watanabe I (1984) Dynamics of nitrogen availability in lowland rice soils: the role of subsoil and effect of moisture regimes. Phil J Crop Sci 9(2): 135–142
144. Von Bulov JWF and Dobereiner J (1975) Potential for nitrogen fixation in maize genotypes in Brazil. Proc Natl Acad Sci USA 72:2389–2393
145. Wada H, Panichsakpatana S, Kimura M and Takai Y (1979) Organic debris as micro-site for nitrogen fixation. Soil Sci Plant Nutr 25(3):453–456
146. Watanabe A and Yamamoto Y (1971) Algal nitrogen fixation in the tropics. Plant and Soil (special volume), 403–413
147. Watanabe I (1978) Biological nitrogen fixation in rice soils. pp 465–478 In Soils and Rice, The International Rice Research Institute, Los Banos, Philippines
148. Watanabe I (1982) *Azolla-Anabaena* symbiosis – its physiology and use in tropical agriculture. pp 169–185. In Microbiology of tropicals soils, Dommergues YR and Diem HG eds, Nijhoff M and Junk W
149. Watanabe I (1984) Use of green manures in Northeast Asia. pp 229–234 In Organic Matter and Rice, The International Rice Research Institute, Los Banos, Philippines
150. Watanabe I and Barraquio WL (1979) Low levels of fixed nitrogen required for isolation of free-living N_2-fixing organisms from rice roots. Nature 277:565–566
151. Watanabe I, Berja NS and Del Rosario DC (1980) Growth of *Azolla* in paddy fields as affected by phosphorus fertilizer. Soil Sci Plant Nutr 26:301-307
152. Watanabe I and Brotonegoro S (1981) Paddy fields. pp 241–263 In Nitrogen Fixation, Vol 1 Ecology, Broughton WJ ed, Clarendon Press, Oxford
153. Watanabe I and Cholitkul W (1982) Nitrogen fixation in acid sulfate paddy soils. Trop Agric Res Ser No 15:219–226
154. Watanabe I, Craswell ET and App AA (1981) Nitrogen cycling in wetland rice fields in south-east and east Asia. pp 4–17 In Nitrogen cycling in south east Asian wet monsoonal ecosystem. Wetselaar R ed, Austral Acad Sci Canberra
155. Watanabe I, De Guzman MR and Cabrera DA (1981) The effect of nitrogen fertilizer on N_2 fixation in paddy field measured by *in situ* acetylene reduction assay. Plant and Soil 49:135–139
156. Watanabe I, De Guzman MR and Cabrera DA (1981) Contribution of basal portion of shoot to N_2 fixation associated with wetland rice. Plant and soil 59:391–398
157. Watanabe I and Espinas C (1985) Relationship between soil phosphorus availability and azolla growth. Soil Sci Plant Nutr 31, (in press)
158. Watanabe I and Lin C (1984) Response of wetland rice to inoculation with *Azospirillum lipoferum* and *Pseudomonas* sp. Soil Sci. Plant Nutri., 30:117–124
159. Watanabe I and Roger PA (1984) Nitrogen fixation in wetland rice fields. pp 237–276 In Current development in biological nitrogen fixation Subba Rao ed Oxford and IBH Pub Co.

160. Wen Qi-Xiao (1984) Utilization of organic materials in rice production in China. pp 45–56 In Organic Matter and Rice, The International Rice Research Institute, Los Banos, Philippines
161. Yamane I (1978) Electrochemical changes in rice soils. pp 381–398 In Soils and Rice, The International Rice Research Institute, Los Banos, Philippines
162. Yoneyama T, Lee KK and Yoshida T (1977) Decomposition of rice straw residues in tropical soils. IV. The effect of rice straw on nitrogen fixation by heterotrophic bacteria in some Philippines soils. Soil Sci Plant Nutr 23:287–295
163. Yoshida S (1981) Fundamentals of rice crop science. 267 p. The International Rice Research Institute, Los Banos, Philippines
164. Yoshida T and Ancajas RR (1971) Nitrogen fixation by bacteria in the root zone of rice. Soil Sci Soc Amer Proc 35:156–157
165. Yoshida T and Rinaudo G (1982) Heterotrophic N_2 fixation in paddy soils. pp 75–107 In Microbiology of Tropical Soils and Plant Productivity, Dommergues YR and Diem HG eds, Martinus Nijhoff and Junk W
166. Yoshida T and Yoneyama T (1980) Atmospheric dinitrogen fixation in the flooded rice rhizosphere as determined by the ^{15}N isotope technique. Soil Sci Plant Nutr 26:551–559
167. Yoshida T, Yoneyama T and Nakajima Y (1983) *In situ* measurement of atmospheric dinitrogen fixation in rice rhizosphere by the N-15 isotope method and acetylene reduction method. Jap J Soil Sci Plant 54:105–108

4. Reappraisal of the significance of ammonia volatilization as an N loss mechanism in flooded rice fields

I R P FILLERY and P L G VLEK

Agro-Economic Division, International Fertilizer Development Center, P.O. Box 2040, Muscle Shoals, Alabama 35662, USA

Key words: N Fertilizer efficiency, N balance, micro-meteorology

Abstract. The role of ammonia volatilization as a nitrogen loss mechanism in lowland rice (*Oryza sativa L.*) has recently been extensively reevaluated using techniques that do not disturb the field environment. This paper summarizes methodologies used in this research and discusses findings from recently conducted micrometeorological studies on ammonia volatilization. Factors affecting ammonia loss and the contribution of this process to the overall nitrogen loss from lowland rice systems are also outlined. Suggestions for future research are included.

Historical perspective

Ammonia volatilization has been generally discounted as an important N loss mechanism in reviews on N transformations in lowland rice systems. The early characterization of flooded soils into separate aerobic and anaerobic zones [42, 43], coupled with the pioneering research on nitrification and denitrification in fallow flooded soils by Shioiri and Mitsui in Japan [40, 42], led many to conclude that nitrification–denitrification was the dominant N loss mechanism in rice [2, 13, 30, 40, 41, 55].

As early as 1935, Screenivasan and Subrahamanyan [45] suggested that NH_3 volatilization could cause high losses of N when urea and dried blood were applied to flooded soils; the role of NH_3 volatilization as an N loss mechanism in flooded alkaline soils was subsequently confirmed by Gupta [27]. However, a succession of NH_3 volatilization studies conducted at the International Rice Research Institute (IRRI), largely with Maahas clay, and research on NH_3 loss in Thailand in 1971 and 1972 [57] failed to demonstrate that NH_3 loss was a major loss mechanism (Table 1). Generally NH_3 volatilization accounted for less than 15% of the N applied as either urea or $(NH_4)_2SO_4$ even though high rates of N were often used (Table 1).

A report by Bouldin and Alimagno [5], which claimed that 40%–60% of the $(NH_4)_2SO_4$-N applied in the field was lost by NH_3 volatilization, rekindled interest in NH_3 volatilization at IRRI and at the International Fertilizer Development Center (IFDC). At about the same time Kissel et al. [33] noted

79

Fertilizer Research 9 (1986) 79–98
© *Martinus Nijhoff/Dr W. Junk Publishers, Dordrecht – Printed in the Netherlands*

80

Table 1. Summary of early studies on ammonia volatilization from flooded soils

Experiment	Estimated ammonia lost % of N applied	Fertilizer material	Fertilizer rate kg N ha^{-1}	Method of measurement	Soil and site
MacRae and Ancajas [37]	11	Urea broadcast	200	Forced draft	Mahas clay, IRRI, laboratory
	8	"	50	"	"
	5	(NH$_4$)$_2$SO$_4$ broadcast	200	"	"
	2	"	50	"	"
	16	Urea broadcast	200	"	Silo silt loam, IRRI, laboratory
	19	"	50	"	"
	3	(NH$_4$)$_2$SO$_4$ broadcast	200	"	"
	7	"	50	"	"
Bouldin and Alimagno [5]	40–60	(NH$_4$)$_2$SO$_4$	100	Open-closed system	Maahas clay, dry season, IRRI
Venture and Yoshida [51]	8.2	Urea broadcast	100	Static chamber with enclosed acid trap	Maahas clay, field, wet season, IRRI
	3.8	(NH$_4$)$_2$SO$_4$ broadcast	100	"	
	3.6	Urea incorporated	100	"	
	1.6	(NH$_4$)$_2$SO$_4$ incorporated	100	"	
Wetselaar et al. [57]	12.4	(NH$_4$)$_2$SO$_4$ broadcast	50	Forced-draft chamber with externally mounted acid trap	Field, dry season, Thailand
	9.2	"	100		
	14.0	(NH$_4$)$_2$SO$_4$ incorporated	50		
Mikkelsen et al. [39]	20	Urea broadcast	90	Forced-draft chamber with externally mounted acid trap	Maahas clay, pot experiment, dry season, IRRI
	15	(NH$_4$)$_2$SO$_4$ broadcast	90		
Mikkelsen et al. [39]	20	(NH$_4$)$_2$SO$_4$ broadcast 10 DT	90	Chamber with enclosed acid trap and natural ventilation	Maahas clay, field, dry season, IRRI
	6	Urea broadcast	90	"	Maahas clay, field, wet season, IRRI
	7	(NH$_4$)$_2$SO$_4$ broadcast	90	"	"

Table 1. contd.

Experiment	Estimated ammonia lost % of N applied	Fertilizer material	Fertilizer rate kg N ha^{-1}	Method of measurement	Soil and site
Vlek and Craswell [52]	50	Urea incorporated	95	Forced-draft with externally mounted acid trap	Decatur and Crowley silt loams, pot, greenhouse
	15	(NH$_4$)$_2$SO$_4$ incorporated	95		
Freney et al. [24]	5	(NH$_4$)$_2$SO$_4$ broadcast and incorporated	80	Micrometeorological	Maahas, field, wet season, IRRI
	11	(NH$_4$)$_2$SO$_4$ topdressed at panicle initiation	40	”	

that NH_3 volatilization was influenced by the rate of air exchange through sampling chambers. The relevance of this finding to floodwater systems was substantiated by Vlek and Stumpe [54] and led Vlek and Graswell [52] to use techniques that maximized the effect of air exchange on NH_3 volatilization. As a result, Vlek and Craswell [52] were able to show that up to 50% of the urea-N applied to puddled Crowley soil was lost via NH_3 volatilization. Vlek and Stumpe [54] found that a lower rate of NH_3 loss (11% N applied) occurred from the same soil when amended with $(NH_4)_2SO_4$ because of the acidification of the floodwater in the course of NH_3 loss. Their findings on NH3 loss from these N sources were consistent with the view that urea was more prone to N loss in flooded soils and, therefore, a poorer source of N for lowland rice. Vlek and Craswell [52] also noted that sulfur-coated urea and deep placement of urea reduced NH_3 fluxes. Both management techniques significantly increased grain yield of rice in the field [49] and pot studies [11].

Mikkelsen et al. [39] also reevaluated the extent of NH_3 volatilization from various N sources applied to flooded rice, albeit with a technique that used low air-exchange rates and therefore probably retarded NH_3 loss. In contrast to Vlek and Craswell [52], these researches found that the NH_3 loss (20% N applied) was similar from urea and $(NH_4)_2SO_4$ when these were applied to floodwater, primarily because alkaline water was used to irrigate experimental plots and pots. Their findings on NH_3 loss from deep-placed urea and sulfur-coated urea were in agreement with those data reported later by Vlek and Craswell [52]. Mikkelsen et al. [49] also confirmed the earlier report by Bouldin and Alimagno [5] that diurnal changes in pH in floodwater was related to diurnal cycles of photosynthesis and respiration in floodwater.

Freney et al. [24] were the first to use micrometeorological techniques to measure NH_3 loss in rice fields. The advantages of this approach over the commonly used methods involving forced-air-exchange chambers are discussed in the next section. Their analyses of NH_3 fluxes after $(NH_4)_2SO_4$ was incorporated to Maahas clay showed NH_3 volatilization to account for about 5% of the N applied. This finding was in excellent agreement with earlier reports of low rates of NH_3 loss from $(NH_4)_2SO_4$ although it was at odds with findings reported by Bouldin and Alimagno [5], also obtained with the Maahas clay but in a different season.

A higher rate of NH_3 volatilization might have been expected in the experiment of Freney et al. [24], especially since ammoniacal N concentrations exceeded $45 \, g/m^3$ in a floodwater system that received alkaline irrigation water. This inconsistency raised questions about the significance of NH_3 volatilization as an N loss mechanism in field environments [24].

The wide disparity in results on NH_3 volatilization from flooded soils prompted IFDC and IRRI to reexamine the importance of this process. Because of the increased use of urea on rice [47] and the higher potential for

NH$_3$ loss from this source, research was focused on this fertilizer. In addition, to determine the practical significance of this process in rice culture, experimental treatments were based on commonly used farmers' management as well as the recommended basal incorporation of N, since studies on constraints to grain yield in the Philippines [12, 26, 31] implied that farmers' N management was often a major cause of poor yield.

Methods used to measure NH$_3$ loss

The range of techniques used to measure NH$_3$ loss is outlined in Table 1. Forced-air-exchange methods that use enclosures and acid traps to absorb NH$_3$ have been commonly employed because of their simplicity. A major limitation of these techniques is their disruptive effect on the natural environment [14]. Thus, while such techniques can be used to establish the potential for NH$_3$ loss, provided that airflow rate does not limit NH$_3$ volatilization, they are not suited to field studies which attempt to quantify the magnitude of this process under undisturbed conditions.

Two attempts have been made to pattern the rate of air exchange through chambers on the wind speed outside the chambers [36, 49]. Conceptually this approach is superior to the single-rate forced-air-exchange methods. However, the use of open-sided chambers, to facilitate rapid air movement over the covered area, does permit the escape of NH$_3$ from the chamber [49]. This characteristic obviously undermines the accuracy of these methods.

Techniques that use analyses of NH$_3$ in air and meteorological measurements such as wind speed, wet- and dry-bulb air temperatures, net radiation, and heat fluxes to estimate NH$_3$ losses (micrometeorological techniques) do not disturb the natural environment and therefore allow the analysis of NH$_3$ loss on a field scale. Three methods have been developed. The energy balance [15, 17] and aerodynamic micrometeorological techniques [34] require that flux measurements be undertaken over large areas of uniformly treated crop (fetches between 150 and 200 m). In contrast, the mass balance or integrated horizontal flux micrometeorological technique [3, 18] can be used on experimental plots with fetches ranging between 20 and 40 m. This feature is especially useful in tropical rice studies, since field size is often small and planted and fertilized areas are discontinous.

The energy balance and aerodynamic micrometerological techniques use derivations of the following expression (equation 1) to calculate NH$_3$ fluxes:

$$F = K \frac{dc}{dz} \qquad (1)$$

where F is the flux of NH$_3$, K is the eddy diffusivity of NH$_3$ in air, c the atmospheric NH$_3$ concentration, and z the height above a crop, soil, or floodwater surface at which measurements of c are undertaken. According to

Denmead [14] c should be measured at heights where the flux of NH_3 is constant with height. This condition can only be satisfied if measurements are undertaken over large areas of uniform crop surface, as noted previously.

The magnitude of the diffusivity of NH_3 in air varies with height, the atmospheric conditions, and the aerodynamic roughness of the surface [14]. Thus, it must be determined during each measurement of F. Because K_{NH_3} is difficult to determine, its value is assumed to be equal to either the eddy diffusivity of water vapor (energy balance technique) or the eddy diffusivity for momentum in air (aereodynamic technique). The more detailed relationships used to calculate NH_3 fluxes according to either the energy balance or the aereodynamic technique and the limitations of each are outlined by Denmead [14].

No assumptions are used in the calculation of NH_3 fluxes by the horizontal flux technique. In this case

$$F = \frac{1}{X} \int_{0_l}^{z} \bar{u}(z)\,\bar{c}(z)\,dz \tag{2}$$

where F is the flux of NH_3, X the fetch, $\bar{u}$ the time-averaged windspeed, $\bar{c}$ the time-averaged atmospheric NH_3 concentration, and z the height at which measurements are taken. A major requirement of this method is that measurements of $\bar{u}$ and $\bar{c}$ should extend above the air layer whose concentration is being modified by the NH_3 emission. According to Denmead [14], a good working approximation is that $z \cong 0.1X$. The accuracy of this method is also dependent on how well the profiles of $\bar{u}$ and $\bar{c}$ are defined experimentally. Denmead [14] suggests that measurements should be made at least at five levels and preferably at more. It is also essential to define the background profile of $\bar{c}$ to calculate the net flux of NH_3 from the treated area. Another requirement is that X, the fetch, be known precisely. This can be easily satisfied by measuring NH_3 fluxes from circular plots, in which case $\bar{u}$ and $\bar{c}$ are measured at the center of the plot while the background $\bar{c}$ is measured in the appropriate upwind quadrant. The radii of circular plots used by researchers have been as follows: 36 m by Beauchamp et al. [3], 25 m by Simpson et al. [46] and Fillery et al. [22], and 22 or 20 m by Fillery et al. [22] and Fillery and De Datta [20]. Ryden and McNeill [44] have thoroughly documented the procedures needed to estimate NH_3 fluxes from rectangular fields.

A major limitation of the mass balance technique is its high labor requirement. A simpler, less laborious method based on the mass balance technique has been proposed [58]. In this case, NH_3 fluxes are estimated from analyses of $\bar{u}$ and $\bar{c}$ made at a single height at which the normalized horizontal flux, $\frac{uc}{F}$, has almost the same value in any atmospheric stability regime. Wilson et al. [58] called this height ZINST. Denmead [14] has presented experimental evidence for ZINST, and the accuracy of the simplified micrometeorological technique has been successfully verified against the mass balance micrometeorological technique in short-grass pasture plots [14] and transplanted rice plots [20]. Two major requirements must be satisfied if the simplified method is to be used. The treated plot should be small (20 m to 50 m radius) and located within a larger plot of uniform crop surface so that the wind profiles are equilibrium ones. However, the technique is useless in fields with well-established canopies because single measurements will not be adequate to predict the strongly modified wind and concentration profiles. In some of the field studies in the Philippines, as much as 50% of the horizontal flux took place within the rice canopy (Bouwmeester and Fillery, in preparation).

Various types of NH_3 traps have been used in micrometeorological studies. Denmead and coworkers [15–18] have routinely used 300 mm long, 20–22 mm ID glass tubes fitted with a central core of wool (190 mm in length), two layers (each 50 mm in length) of 3-mm diameter glass beads, and a cap of glass wool at both ends (each 25 mm in length). Phosphoric acid (10 ml of 1%) is introduced to the inlet end; the tubes are inverted and air drawn through at $10 \, L \, min^{-1}$ for 2 hours during the daytime and at $5 \, L \, min^{-1}$ in overnight measurements. Ammonium-N is recovered by back leaching each trap with deionized distilled H_2O until 20 ml of leachate is collected. The acid traps are then treated with $0.1 \, N$ NaOH and leached with additional deionized H_2O, after which they are dried in a convection oven before reuse. A modification of this method was introduced by Fern [19] who used glass tubes impregnated with citric acid to trap atmospheric NH_3.

Another technique is to use 250 ml suction flasks filled with $190 \, cm^3$ of 3 mm diameter glass beads and 50 ml of 3% (vol/vol) H_3PO_4 solution. In this case, samples of acid are withdrawn at the start and at the end of each sampling period, and the acid is replaced between each run. The efficiency of entrapment of NH_3 must be determined however [3].

Harper et al. [29] and McInnes [38] trapped atmospheric NH_3 in gas scrubbing bottles filled with sulfuric acid. While the acid can be easily recovered for analysis with this approach and the washing procedures are less laborious than those needed for the glass wool-glass bead traps, the flow rate of air through such traps can be restricted by trap dimensions and volume of acid added. Thus, less NH_3 can be be trapped with these procedures over a given sampling period. This effect probably will not reduce the accuracy of NH_3 analysis where large fluxes of NH_3 occur. However, there is a larger uncertainty regarding the analysis of NH_3 fluxes from background areas or the determinations of low rates of fluxes from treated areas, especially if relatively insensitive analytical techniques are used to measure NH_3^+.

Numerous analytical techniques have been used to measure the content of NH_3 in acid traps. Harper et al. [29] and McInnes [38] employed automated indophenol methods [56, 50]. Turner et al. [48] recovered NH_3 from citric acid-impregnated glass tubes by using 2% Nessler's reagent. Color development was then determined by spectrophotometry. Abbas and Tanner [1] report that a fluorescence method based on the derivatization of NH_3 by o-phthalaldehyde permitted real-time analyses of NH_3 in air. This method has not been adapted to NH_3 volatilization studies. Denmead et al. [15–17], Fillery and De Datta [20], Fillery et al. [22], and Simpson et al. [46] used NH_3 specific ion electrodes. This technique permitted NH_4^+-N analysis to be undertaken at the experimental site soon after the recovery of NH_4^+-N from acid traps. Automated indophenol methods are typically less sensitive and more difficult to adapt to field use than procedures that use NH_3 specific ion electrodes. However, the manual operation coupled with the rigorous calibration requirements associated with specific ion electrodes makes this particular technique laborious. The choice of analytical technique will obviously be dictated by facilities and manpower available.

Factors that affect NH_3 volatilization

Recent reviews on NH_3 volatilization [53, 25] include comprehensive treatises on factors that affect this process. Our paper will concentrate on findings from recently conducted field studies.

Fertilizer management, through its influence on the concentration of ammoniacal N in the floodwater, has a pronounced effect on the overall NH_3 loss [22]. In Philippine studies the highest sustained concentrations of ammoniacal N have been detected when urea or $(NH_4)_2SO_4$ was applied to floodwater 2–4 weeks after the transplanting of rice seedlings [22]. High concentrations of ammoniacal N can also form in floodwater when urea is applied to 10–20 mm of floodwater before incorporation is attempted [8, 9]. Incorporation of urea to puddled soil, when it is undertaken without standing floodwater, can reduce the concentrations of ammoniacal N in floodwater [22].

Urease activity, cation exchange, and immobilization of N in soil are processes that could potentially affect the concentration of ammoniacal N in floodwater and indirectly the extent of NH_3 loss [25]. Assimilation of NH_4^+-N by algae, weeds, and rice plants could also decrease the quantity of ammoniacal N available for NH_3 loss. For example, significant quantities of ^{15}N (15% N applied) have been detected in algae and weeds 2 weeks after the application of urea in some field studies [10]. The competitiveness of the rice crop for NH_4^+-N varies with growth stage. About 40% of ^{15}N-labeled urea applied to a rice crop at panicle initiation was recovered in aboveground plant material 10 days after N was applied in studies conducted by Fillery et al. [21]. In contrast, after a similar period of uptake, rice plants generally contained less than 10% of the ^{15}N-labeled urea that was applied to floodwater between 2 and 3 weeks after transplanting (AT) and negligible quantities of ^{15}N ($\cong 1\%$ N applied) 1 week after the broadcast and incorporation (BI) of urea [21].

The quantity of ammoniacal N in floodwater is an index of the potential for NH_3 volatilization. The rate of NH_3 loss is dependent on the equilibrium vapor pressure of NH_3 (P_{NH_3}) in floodwater and on wind speed (Figure 1).

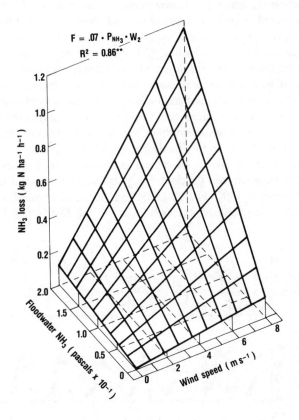

Figure 1. Effect of partial pressure of NH_3 (P_{NH_3}) and wind speed (W_2) on NH_3 fluxes in AT studies conducted at Muñoz and Los Baños, 1982, W_2 is the wind speed at a height of 1.2 m.

The vapor pressure of NH_3 floodwater is a function of the ammoniacal N concentration, pH, and temperature [28] and can be estimated by using the following relationships:

$$P_{NH_3} = \frac{NH_3 \, (aq)}{K_H} = \frac{K \cdot AN}{K_H \, (H + K)} \qquad \begin{array}{l} \text{with } PK_H^\circ = 1.77 \\ PK^\circ = 9.24 \end{array}$$

where K for nonstandard state temperatures can be obtained from Beutier and Renon [4]:

$$\ln K = -177.95 + 1{,}843.22/T + 31.4335 \ln T - 0.0545 \, T \quad (4)$$

while for K_H a similar correction was reported:

$$\ln K_H = 160.559 - 8,621.06/T - 25.6767 \ln T + 0.035388 \, T$$

$$(5)$$

Other functions for temperature corrections have been reported [28, 32] that are equally useful if properly reparameterized. Pressures and concentrations in Equation (3) are expressed in atmospheres and moles L^{-1}, respectively.

According to Vlek and Craswell [53], the content of aqueous NH_3 in floodwater increases about tenfold per unit increase in pH in the pH range 7.5 to 9.0, and it increases approximately linearly with increasing temperature at a given total concentration of ammoniacal N. In field systems the pH in the floodwater displays a diurnal pattern (Figure 2) that appears to be synchronized with the cycles of photosynthesis and net respiration or the depletion and the addition of CO_2 to the floodwater [5, 39, 22]. The magnitude of the diurnal fluctuations in pH in floodwater is influenced by fertilizer management and the stage of crop development, presumably because of their effect on the algal biomass [39]. For example, minimum and maximum pH values of 7.8 and 8.6, respectively, typically occur in the floodwater 2–3 weeks

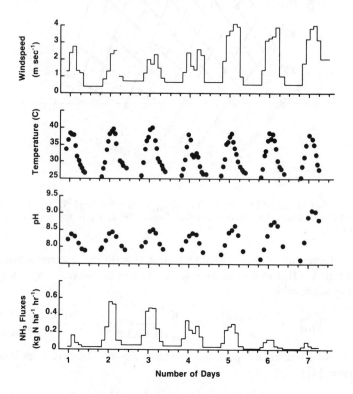

Figure 2. Diurnal pattern of NH_3 fluxes, wind speed, pH, and temperature in floodwater during an AT study conducted at Los Baños. Adapted from Fillery et al. [22].

after transplanting. Progressively higher maximum pH values (up to pH 9.7) occur after urea is basally incorporated to soil largely because P is also applied at this time and, as a result, the algal biomass increases rapidly [22]. Small diurnal fluctuations in pH (minima and maxima of pH 7.8 and 8.2, respectively) are normally detected at panicle initiation when the rice crop shades the floodwater [22].

The release of NH_4^+-N and HCO_3^- into poorly buffered floodwater during urea hydrolysis can cause the pH to be maintained at about 8.0 [53]. Depending on the biological activity and the strength of the NH_4HCO_3 buffer, this pH may be maintained for a few days. For example, Vlek and Craswell (53) observed that diurnal pH fluctuations were delayed by 48 hours when urea was applied to floodwater in a field study in Alabama, United States. This effect was not observed in Philippine studies, possibly because of the higher biological activity in floodwater and the formation of comparatively lower concentrations of ammoniacal N $(14\,mg\,L^{-1})$ in floodwater following the application of urea. However, Fillery et al. [21] concluded that a solution of NH_4HCO_3 buffered the floodwater pH at about 8.0 for 2 days following the application of $(NH_4)_2SO_4$ to floodwater that contained between $3-4\,meq\,L^{-1}$ alkalinity. In this case ammoniacal N concentrations in floodwater $(30-40\,mg\,N\,L^{-1})$ were initially within the range measured by Vlek and Craswell [53] in their field study.

Wind speed or air exchange is another environmental parameter that markedly affects NH_3 volatilization from floodwater [6, 54]. In field studies conducted in the Philippines, ammonia volatilization increased linearly as wind speed increased (Figure 1). As would be expected on the basis of equation [2], analyses of the relationship between P_{NH_3}, wind speed, and NH_3 volatilization revealed that there was a good fit $(R^2 = 0.90)$ to the following relationship between the partial pressure of NH_3 (P_{NH_3}) in the floodwater and wind speed at 1.2 meters (W_2) during the volatilization process:

$$F = k \cdot P_{NH_3} \cdot W_2 \tag{6}$$

where the constant k incorporates a factor for the fetch (l/X of equation 2) as well as factors relating P_{NH_3} and W_2 to the concentration and wind profiles, respectively. Such findings highlight the importance of wind speed and clearly demonstrate why low losses of NH_3 were frequently recorded in studies that used low-air-exchange, forced-draft sampling techniques, even though high rates of N application were often used in these studies (Table 1).

The effects of changing ammoniacal N concentration, pH, temperture, and wind speed have been evaluated simultaneously by using an ammonia diffusion model and data from wind tunnel studies on NH_3 volatilization [6]. This model has been recently refined on the basis of information obtained in the field in the Philippines (Figure 3). Essentially the updated model predicts higher fluxes of NH_3 for given levels of ammoniacal N, pH, temperature, and wind speed than were reported previously [6].

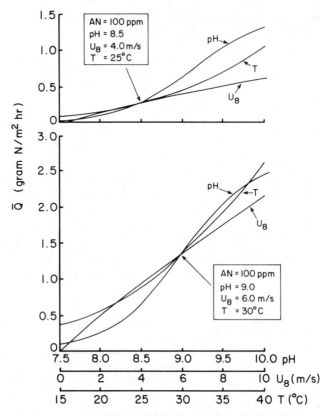

Figure 3. Variation of ammonia volatilization with wind speed U_8, temperature (T), and pH as predicted using a diffusion model and field data. (Bouwmeester and Fillery, unpublished data.)

Ammonia volatilization is an inherently acidifying process as illustrated in the following equation.

$$NH_4^+ \rightleftharpoons NH_3 + H^+ \tag{7}$$

Alkalinity (chiefly HCO_3) must therefore be present in floodwater to buffer the production of H^+ if NH_3 volatilization is to be sustained [54]. Urea hydrolysis is considered to be a major source of bicarbonates in floodwater that is amended with urea, especially floodwater that is naturally low in alkalinity [53]. Irrigation water is probably the major source of alkalinity in floodwater systems amended with N forms such as $(NH_4)_2SO_4$ or $(NH_4)_2HPO_4$. The quantities of alkalinity that are needed in irrigation water to sustain different rates of NH_3 losses from NH_4^+-N sources are shown in Figure 4. According to this relationship, the irrigation of flooded soils with water that is relatively low in alkalinity could result in appreciable losses of NH_3 provided NH_4^+-N sources are applied at a stage when evapotranspiration and repeated application of irrigation water have increased the content

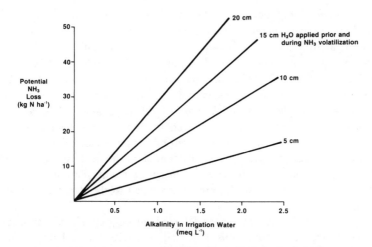

Figure 4. Relationship between potential NH_3 loss, alkalinity in irrigation water, and quantity of water supplied to flooded rice fields.

of alkalinity in the floodwater. For example, the addition over a period of 2–3 weeks of 10 cm of irrigation water containing on average 1.6 meq L^{-1} total alkalinity could theoretically result in an NH_3 loss of 22 kg N ha^{-1}. This value agrees very well with the NH_3 loss from $(NH_4)_2SO_4$ (21.5 kg N ha^{-1}) from a field system that received approximately 10 cm of irrigation water (1.6 meq L^{-1}) between the transplanting of rice seedlings and the application of $(NH_4)_2SO_4$ [23]. Total titratable alkalinity in floodwater has been shown to display a diurnal pattern, especially after the application of $(NH_4)_2SO_4$ [23]. We surmise that the formation of OH^- during periods of increased soil reduction at night (Equation 8) coupled with net CO_2 production contributed HCO_3^- to the floodwater (Equation 9).

$$Fe(OH)_3 + e^- \rightleftharpoons Fe^{2+} + 3OH^- \qquad (8)$$

$$3CO_2 + 3OH^- \rightleftharpoons 3HCO_3^- \qquad (9)$$

Confirmation of this mechanism poses as interesting challenge to soil scientists working in flooded soil systems. Such findings suggest that the inherent total titratable alkalinity in many irrigated rice systems, especially in the dry season, may be sufficient to induce modest rates of NH_3 loss from $(NH_4)_2SO_4$.

Ammonia volatilization and N management

The effect of N management on NH_3 volatilization from lowland rice fields has recently been examined in detail in transplanted rice in the Philippines [22, 20] and in direct-seeded rice in New South Wales, Australia [46]. In every case the integrated horizontal or mass balance micrometeorological technique was used to measure NH_3 fluxes. Although these environments may not necessarily be representative of rice production worldwide, the results are reviewed here to demonstrate the usefulness of the methodology and illustrate the role NH_3 loss can play in rice production.

N management used in the Philippine studies followed both recommended and farmers' practices. Urea was incorporated into puddled soil immediately before the transplanting of rice seedlings (recommended management), applied to floodwater 14–21 days after transplanting (a common farmer practice), or topdressed to floodwater at panicle initiation (recommended management). Concurrent micrometeorological measurements of NH_3 volatilization have also been attempted in the field using $(NH_4)_2SO_4$, urea, and urea amended with the urease inhibitor, phenyl phosphorodiamidate (PPD). In the Australian study, urea was applied to the floodwater 8 weeks after the rice crop was direct seeded.

In general, NH_3 fluxes have been detected immediately after the application of $(NH_4)_2SO_4$ to flooded soils [24, 20] and within 2–4 hours after the application of urea [22, 20, 46]. Thereafter, NH_3 fluxes have displayed a diurnal pattern (Figure 2) that is synchronized with the diurnal fluctuations in pH and temperature in floodwater and diurnal fluctuations in wind speed (Figure 2). The maximum NH_3 fluxes are generally observed immediately following the application of $(NH_4)_2SO_4$ [24, 20] or 2–3 days after the application of urea [24, 20]. Ammonia volatilization typically ceases about 7–10 days after fertilizer application, depending on the N source and management [24, 22, 46, 20].

The highest losses of NH_3 have been detected in the Philippine studies when urea was applied to the floodwater 2–4 weeks after the transplanting of rice seedlings. For example, 47% of the urea-N topdressed to floodwater in a study at Muñoz in the 1982 dry season was volatilized as NH_3. A lower rate of NH_3 loss (27% N applied) was detected in a similar experiment conducted at Los Baños in the same season, probably because wind speed was lower at Los Baños than at Muñoz [22]. Comparable dry season studies conducted at Muñoz at 1983 and at Los Bañoz in 1984 showed that 36% and 53%, respectively, of the urea N applied was lost by NH_3 volatilization [20], (Fillery and Byrnes, unpublished data). These findings highlight the importance of NH_3 volatilization as an N loss mechanism in flooded rice fields.

Substantially lower NH_3 losses (13%–15% N applied) were detected after the broadcast and incorporation of urea into puddled soil which had been drained of floodwater before the application of N, chiefly because ammoniacal N concentrations in floodwater were well below those observed after urea was applied directly to floodwater [22]. Low rates of NH_3 loss (10%–13% N applied) were also detected after urea was applied to flooded rice at panicle initiation [22]. Aside from N uptake, it appeared that the rice crop shaded the floodwater and thereby suppressed the increase in pH attributed to photosynthetic activity in floodwater. Alternatively, the denser plant canopy at panicle initiation may have restricted air exchange at the flood-water-air interface [22].

Simpson et al. [46] reported that a small loss of NH_3 (11% N applied) occurred after the application of urea to an 8-week-old direct-seeded rice crop despite the presence of high pH values and elevated urea concentrations in floodwater and high wind speeds. Low urease activity at the soil-floodwater interface and the depth of floodwater ($\cong 0.15$ m) may have retarded the accumulation of ammoniacal N in floodwater in this study [46]. Another explanation for the low NH_3 loss is that a temperature inversion in flood-water, induced by the adsorption of short-waved radiation in turbid water, restricted mechanical and buoyancy mixing forces, and consequently the transport of NH_3 to the floodwater surface [35]. Clear, shallow floodwater prevailed in the Philippine studies discussed earlier.

Although the role of NH_3 volatilization as an N loss mechanism from urea is now well established in lowland rice, the importance of this mechanism after the application of $(NH_4)_2SO_4$ is less well understood. Fillery and De Datta [20] demonstrated that NH_3 volatilization (38% N applied) was similar from $(NH_4)_2SO_4$ and urea when these sources were applied to lowland rice 18 days after the transplanting of rice seedlings. Their report of a high NH_3 loss from $(NH_4)_2SO_4$ confirms the results obtained by Bouldin and Alimagno [5] but is at odds with data obtained by Freney et al. [24] and Wetselaar et al. [57]. The role of alkalinity in the NH_3 volatilization process was discussed earlier. Sufficient alkalinity was present in the floodwater at the time of the application of the $(NH_4)_2SO_4$ in the Moñoz study to account for the NH_3 fluxes [21]. The possible sources of this alkalinity were outlined earlier. The important question to be addressed now is whether the levels of alkalinity detected in floodwater in this study are typical for irrigated Asian rice fields.

Contribution of NH_3 loss to the total N loss

To fully understand the significance of NH_3 volatilization as an N loss mechanism, it is important to determine its contribution to the overall N loss after N fertilization. This can be achieved by conducting ^{15}N loss measurements concurrently with micrometeorological studies of NH_3 fluxes. In this case it is assumed that the N loss from microplots is similar to that which occurs in the larger field plot.

In the Australian study discussed earlier [46], NH_3 fluxes accounted for 24% of the ^{15}N lost. These researchers attributed the balance of the ^{15}N loss to nitrification-denitrification. Fillery et al. [23] and Fillery and De Datta [20] report that NH_3 volatilization accounted for about 92%–100% of the ^{15}N unaccounted for when either urea or $(NH_4)_2SO_4$ was applied as AT treatments at Muñoz, Phillipines. In contrast, at Los Baños in the 1982 dry season, NH_3 fluxes from AT and BI treatments accounted for 45% and 50% of the

94

Table 2. Contribution of ammonia volatilization to the total N loss in flooded rice fields

Experiment	Year	Site[a]	N source	Treatment[b]	Total ^{15}N loss	NH₃ Volatilized	Ratio of NH₃ loss to total ^{15}N loss
					$-$ % N applied $-$		
Fillery et al. [23]	1981	Muñoz	Urea	BI	18	15	0.83
Fillery et al. [23]	1982	Los Baños	Urea	BI	26	13	0.50
Fillery et al. [23]	1982	Muñoz	Urea	AT	45	47	1.02
Fillery et al. [23]	1982	Los Baños	Urea	AT	60	27	0.45
Fillery and DeDatta [20]	1983	Muñoz	Urea	AT	42	36	0.86
Fillery and DeDatta [20]	1983	Muñoz	$(NH_4)_2SO_4$	AT	44	38	0.86
Fillery and Byrnes (unpublished data)	1984	Los Baños	Urea	AT	54	53	0.98
Simpson et al. [46]	1980	Griffith	Urea	DS	45	11	0.24

[a]Muñoz and Los Baños are in the Philippines; Griffith is in New South Wales, Australia.
[b]BI is broadcast and incorporated; AT is applied to floodwater 2 to 4 weeks after transplanting; and DS applied to flooded direct-seeded rice.

[15]N lost, respectively (Table 2). Again it was assumed that nitrification-denitrification accounted for the balance of the ^{15}N loss [23]. A subsequent AT study conducted at Los Baños in 1984 showed that NH_3 volatilization was the dominant loss mechanism (Table 2). Direct measurements of N_2O and [15]N_2 fluxes accounted for less than 1% of the ^{15}N-labeled urea that was applied (Fillery and Byrnes, unpublished data).

Similar pH values and concentrations of ammoniacal N were detected in the floodwater in the AT studies conducted in 1982 [21]. However, wind speed and temperature in floodwater, in particular, differed between these AT studies. Fillery et al. [23] suggest that the high wind speeds at Muñoz may have promoted NH_3 loss and probably precluded any important N loss via nitrification-denitrification. Conversely, the low wind speeds at Los Baños in 1982 possibly restricted NH_3 volatilization and thereby increased the potential for N loss via nitrification-denitrification. The higher temperatures at Los Baños than at Muñoz may have also influenced the relative contribution of each gaseous loss mechanism to the total N loss. Soil properties also differed between Muñoz and Los Baños. However, the contrasting results obtained in the Los Baños AT studies conducted in 1982 and 1984 on similar soil (Table 2) tend to minimize the importance of soil properties as factors affecting the relative contribution of NH_3 loss to the total N loss.

Overall these findings indicate that neither NH_3 volatilization nor nitrification-denitrification is an exclusive loss mechanism in lowland rice fields. The similarity in the total ^{15}N loss from AT treatments (Table 2), irrespecitve of the extent of NH_3 volatilization, suggests that NH_3 loss and nitrification-denitrification may be complementary loss mechanisms in flooded soils.

Future research needs

Much of the information on field NH_3 fluxes discussed here was collected at only three locations: two in the Philippines and the other in Australia. It is important that the role of NH_3 volatilization be examined in other rice-growing areas, in direct-seeded as well as transplanted rice, on contrasting soils, and under different environmental conditions. Wherever possible such measurements should be undertaken concurrently with ^{15}N loss measurements to substanitate the information on the contribution of NH_3 volatilization to the total N loss.

A simplified micrometeorological technique, proposed by Wilson et al. [58], appears to accurately estimate NH_3 fluxes. Further simplification of the techniques used to measure atmospheric NH_3 concentrations is needed to facilitate measurement of NH_3 fluxes at locations that cannot support current measurement techniques. Analyses of alkalinity and pH in irrigation water as well as in floodwater and the measurement of wind speed would also provide invaluable data on the potential for NH_3 volatilization.

Preliminary information suggests that NH_3 volatilization and nitrification-denitrification may be complementary loss mechanisms in lowland soils. Further research is needed to substantiate this possibility, and the soil and atmospheric conditions that affect the relative contribution of either loss mechanism to the overall N loss.

Initial research on the urease inhibitor, phenyl phosphorodiamidate, (PPD) has shown that the urease inhibitor can substantially reduce the concentrations of ammoniacal N in floodwater in AT treatments [22] and reduce NH_3 volatilization [7, 20]. However, PPD appears to decompose rapidly in flooded soil. More effective urease inhibitors are needed to completely eliminate the accumulation of ammoniacal N in floodwater and thereby drastically reduce the potential for NH_3 volatilization after the application of urea.

References

1. Abbas R and Tanner RL (1981) Continuous determination of gaseous ammonia in the ambient atmosphere using fluorescence derivatization. Atmos Environ 15:277–281
2. Abichandani CT and Patnaik S (1955) Mineralizing action of lime on soil nitrogen in waterlogged rice soils. Int Rice Comm News Letter 13:11–13
3. Beauchamp EG, Kidd GE, and Thurtell G (1978) Ammonia volatilization from sewage sludge applied in the field. J Environ Qual 7:141–146
4. Beutier D and Renon H (1978) Representation of NH_3-H_2S-H_2O, NH_3-CO_2-H_2O, and NH_3-SO_2-H_2O vapor-liquid equilibria. Ind Chem Process Des Dev 17:220–230
5. Bouldin DR and Alimagno BV (1976) volatilization losses from IRRI paddies following broadcast applications of fertilizer nitrogen. International Rice Research Institute. Internal Report. P.O. Box 933. Manila, Philippines (Unpublished mimeo)
6. Bouwmeester RJB and Vlek PLG (1981) Rate control of ammonia volatilization from rice paddies. Atmos. Environ. 15:131–140
7. Byrnes BH, Savant NK, and Craswell ET (1983) Effect of a urease inhibitor, phenyl phosphorodiamidate, on the efficiency of urea applied to rice. Soil Sci Soc Am J 47:270–274
8. Cao Zhi-Hong, De Datta SK, and Fillery IRP (1984) Effect of placement methods on floodwater properties and recovery of applied nitrogen (^{15}N-labelled urea) in wetland rice. Soil Sci Soc Am J 48:196–203
9. Craswell ET, De Datta SK, Obcemea WN, and Hartantyo M (1981) Time and mode of nitrogen fertilizer application to tropical wetland rice. Fert Res 2:247–259
10. Craswell ET, De Datta SK, Weeraratne CS, and Vlek PLG (1985) Fate and efficiency of nitrogen fertilizers applied to wetland rice. I Philippines Fert Res 6:49–63
11. Craswell ET and Vlek PLG (1979) Greenhouse evaluation of nitrogen fertilizers for rice. Soil Sci Soc Am J 44:1884–1888
12. De Datta SK, Garcia FV, Abilay WP Jr., Alacantara JM, Mandac A, and Marciano VP (1979) Constraints to high rice yields, Nueva Ecija, Philippines. In Farm-level constraints to high rice yields in Asia: 1974–77, Int Rice Inst, Los Baños, Laguna, Philippines, pp. 191–234
13. De Datta SK and Zarata PM (1970) Environmental conditions affecting the growth characteristics, nitrogen response, and grain yield of tropical rice. Biometeorology 4:71–89
14. Denmead OT (1983) Micrometeorological methods for measuring gaseous losses of nitrogen in the field. In J R Freney and J R Simpson (Eds) Gaseous loss of nitrogen from plant-soil systems, Martinus Nijhoff/Dr W. Junk Publishers, The Hague, pp. 133–157

15. Denmead OT, Freney JR, and Simpson JR (1976) A closed ammonia cycle within a plant canopy. Soil Biol Biochem 8:161–164
16. Denmead OT, Freney JR, and Simpson JR (1982) Dynamics of ammonia volatilization during furrow irrigation of maize. Soil Sci Soc Am J 46:149–155
17. Denmead OT, Simpson JR, and Freney JR (1974) Ammonia flux into the atmosphere from a grazed pasture. Nature 185:609–610
18. Denmead OT, Simpson JR, and Freney JR (1977) The direct field measurement of ammonia emission after injection of anhydrous ammonia. Soil Sci Soc Am J 41: 1001–1004
19. Fern M (1979) Method for determination of atmospheric ammonia. Atmos Environ 13:1385–1393
20. Fillery IRP and De Datta SK (1985) Effect of N source and a urease inhibitor on NH_3 loss from flooded rice (1) Micrometeorological technique, ammonia fluxes and ^{15}N loss. Soil Sci Soc Am J (in press)
21. Fillery IRP, Roger PA, and De Datta SK (1985) Effect of N source and a urease inhibitor on NH_3 loss from flooded rice (2). Floodwater properties and submerged photosynthetic biomass. Soil Sci Soc Am J 49 (in press)
22. Fillery IRP, Simpson JR, and De Datta SK (1984) Influence of field environment and fertilizer management of ammonia loss from flooded rice. Soil Sci Soc Am J 48:914–920
23. Fillery IRP, Simpson JR, and De Datta SK (1985) Contribution of ammonia volatilization to total N loss after applications of urea to wetland rice fields. Fert Res (in press)
24. Freney JR, Denmead OT, Watanabe I, and Craswell ET (1981) Ammonia and nitrous oxide losses following applications of ammonium sulphate to flooded rice. Aust J Agric Res 32:37–45
25. Freney JR, Simpson JR, and Denmead OT (1983) Volatilization of ammonia. In J R Freney and J R Simpson (Eds) Gaseous loss of nitrogen from plant-soil systems, Martinus Nijhoff/Dr W. Junk Publishers, The Hague, pp. 133–157
26. Garcia FV, Abilay WP Jr., Alcantara JM, and De Datta SK (1982) Yield response of wetland rice to various fertilizer management practices and to other inputs in farmers' fields in the Philippines. Philippine J Crop Sci 7(1):51–58
27. Gupta SP (1955) Loss of nitrogen in the form of ammonia from water-logged paddy soil. J Indian Soc Soil Sci 3:29–32
28. Hales JM and Drewes DR (1979) Solobility of ammonia in water at low concentrations. Atmos Environ 13:1133–1147
29. Harper LA, Catchpoole VR, Davis R, and Weier KL (1983) Ammonia volatilization: soil, plant, and microclimate effects on diurnal and seasonal fluctuations. Agron J 75:212–218
30. International Atomic Energy Agency (1966) IAEA Tech. Rep. Ser. 55.
31. IRRI (1980) Annual Report for 1979. International Rice Research Institute. P.O. Box 933. Manila, Philippines
32. Johansson O and Wedborg M (1980) The ammonia-ammonium equilibrium in seawater at temperatures between 5 and 25 °C. J Solution Chem 9:37–44
33. Kissel DE, Brewer HL, and Arkin GF (1977) Design and test of a field sampler for ammonia volatilization. Soil Sci Soc Am J 41:1133–1138
34. Lemon E and van Houtte R (1980) Ammonia exchange at the land surface. Agron J 72:876–883
35. Leuning R, Denmead OT, Simpson JR, and Freney JR (1985) Dynamics of ammonia loss from shallow floodwater. Atmos Environ (page)
36. Lockyer DR (1984) A system for the measurement in the field of losses of ammonia through volatilization. J Sci Food Agric 35:837–848
37. MacRae IC and Ancajas R (1970) Volatilization of ammonia from submerged tropical soils. Plant and Soil 33:97–103
38. McInnes KJ (1985) Aspects of ammonia volatilization from surface applied urea fertilizers. PhD dissertation, Kansa State University, Manhattan, Kansas
39. Mikkelsen DS, De Datta SK, and Obcemea (1978) Ammonia volatilization losses from flooded rice soils. Soil Sci Soc Am J 42:725–730

40. Mitsui S (1954) Inorganic nutrition, fertilization and soil amelioration for lowland rice. Yokendo Ltd. Tokyo, Japan
41. Mitsui S (1977) Recognition of the importance of denitrification and its impact on various improved mechanized applications of nitrogen to rice plants. In Proc. Int. Symp. on soil environment and fertility management in intensive agriculture (SEFMIA). Tokyo, Japan, pp. 259–268
42. Pearsall WH (1938) The soil complex in relation to plant communities. I. Oxidation-reduction potentials in soils. J Ecol 26:180–193
43. Pearsall WH and Mortimer CH (1939) Oxidation reduction potentials in waterlogged soils, natural water and muds. J Ecol 26:180–193
44. Ryden JC and McNeill JE (1984) Application of the micrometeorological mass balance method to the determination of ammonia loss from a grazed sward. J Sci Food Agric 35:1297–1310
45. Screenivasan A and Subrahamanyan V (1935) Biochemistry of waterlogged soils. II. Carbon and nitrogen transformations. J Agric Sci 25:6–21
46. Simpson JR, Freney JR, Wetselaar R, Muirhead WA, Leuning R, and Denmead OT (1984) Transformations and losses of urea nitrogen after application to flooded rice. Aust J Agric Res 35:189–200
47. Stangel PJ (1979) Nitrogen requirement and adequacy of supply for rice production. In Nitrogen and rice. Inernational Rice Research Institute. P.O. Box 933. Manila, Philippines. pp. 45–69
48. Turner RD, Sirovica S, and Bouwmeester RJB (1983) A sensitive colorimetric method of Civil and Sanitary Engineering. Michigan State University. East Lansing, Michigan
49. Vallis I, Harper LA, Catchpoole VR, and Weier KL (1982) Volatilization of ammonia from urine patches in a subtropical pasture. Aust J Agric Res 33:97–107
50. Varley JA (1966) Automatic methods for the determination of nitrogen, phosphorus, and potassium in plant material. Analyst 91:119–126
51. Ventura WB and Yoshida T (1977) Ammonia volatilization from a flooded tropical soil. Plant and Soil 46:521–531
52. Vlek PLG and Craswell ET (1979) Effect of N source and management on ammonia volatilization losses from flooded rice-soil systems. Soil Sci Soc Am J 43:352–358
53. Vlek PLG and Craswell ET (1981) Ammonia volatilization from flooded soils. Fert Res 2:227–245
54. Vlek PLG and Stumpe JM (1978) Effects of solution chemistry and environmental conditions on ammonia volatilization losses from aqueous systems. Soil Sci Soc Am J 42:416–421
55. Watanabe I and Mitsui S (1979) Denitrification loss of fertilizer nitrogen in paddy soils – its recognition and impact. IRRI Research Paper Series 37:1–10
56. Weier KL, Catchpoole VR, and Harper LA (1980) An automated colorimetric method for the determination of small concentrations of ammonia in air. Div. Trop. Crops and Past. CSIRO, Australia Tropic Agron Tech Mem No. 20
57. Wetselaar RT, Shaw T, Firth P, Oupatham J, and Thitiopoca H (1977) Ammonia volatilization from variously placed ammonium sulphate under lowland rice field conditions in central Thailand. Proc. Int. Seminar SEFMIA. October 10–14, 1977. Tokyo, Japn. Soc of Sci of Soil and Manure. Tokyo, Japan
58. Wilson JR, Thurtell GW, Kidd GE, and Beauchamp EG (1982) Estimation of the rate of gaseous mass transfer from aa surface plot to the atmosphere. Atmos Environ 16:1–7

5. Denitrification losses in flooded rice fields*

KR REDDY[1] and WH PATRICK Jr[2]

[1] Professor, University of Florida, Institute of Food and Agricultural Sciences, Central Florida Research and Education Center, P.O. Box 909, Sanford, FL 32771 and [2] Boyd Professor, Center for Wetland Resources, Louisiana State University, Baton Rouge, LA 70803, USA

Key words: nitrification, nitrogen loss, lowland soils, paddy field, waterlogged soil

Introduction

Nitrogen transformations in a flooded rice soil are much the same as N transformations in drained soil systems, though the special soil/environmental conditions prevailing in flooded rice soils alter the rate at which these processes occur. The key N transformations in a flooded soil system include (1) mineralization of organic N; (2) nitrification of NH_4^+; (3) NH_3 volatilization; (4) denitrification; and (5) N_2 fixation (Figures 1a, b). Agronomic and ecological significance of these processes in the gain or loss of N from a flooded soil and sediment system has been studied by several researchers. The objective of this paper is to review recent research findings on the significance of nitrification-denitrification in flooded rice soils.

Soil/environmental conditions

Flooding of a soil results in displacement of soil O_2 with water, with any dissolved O_2 present in the pore water being readily consumed during microbial respiration, thus making soil profile devoid of O_2 [29, 30, 7, 65]. Supply of O_2 to the flooded soil is renewed in two ways: i.e. (1) diffusion of O_2 through the overlying floodwater and consumption at the soil-water interface, and (2) transport of O_2 through the stems of rice and other wetland plants to the roots and subsequent diffusion of O_2 into the rhizosphere. The greater potential consumption of O_2 compared to the renewal rate results in the development of two distinct soil layers: (1) an oxidized or aerobic soil layer which ranges from a few millimeters in thickness in soils of high microbial activity to 1 to 2 cm in soils of low biological activity, and (2) an underlying reduced or anaerobic soil layer in which no free

---

*Joint contribution from the University of Florida and Louisiana State University. Florida Agricultural Experiment Stations Journal Series No. 5997.

Fertilizer Research 9 (1986) 99–116
© *Martinus Nijhoff/Dr W. Junk Publishers, Dordrecht – Printed in the Netherlands*

Flooded Soil

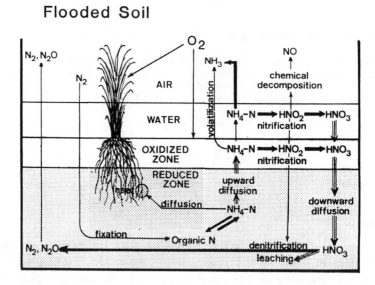

Figure 1(a). Schematic presentation of the N transformations functioning in the oxidized and reduced soil layers of flooded lowland soil.

Rhizosphere

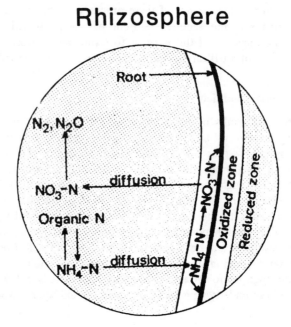

Figure 1(b). Schematic presentation of the N transformations functioning in the oxidized and reduced soil layers of rice rhizosphere.

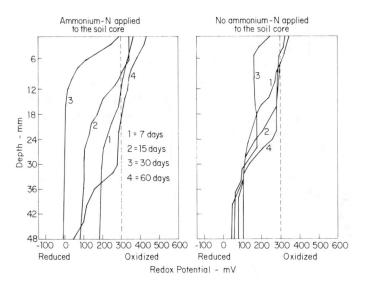

Figure 2. Thickness of oxidized soil layer of a flooded Crowley silt loam soil treated with ammonium fertilizers.

O_2 is present (Figure 1a). This profile differentiation has been characterized for flooded soils and for lake and ocean sediments, by several research workers [58, 42, 1, 22, 34].

The thickness of the oxidized soil layer can be characterized by measuring the oxidation-reduction potential of the soil profile. It is well established by earlier research [32] that O_2 disappears from a flooded soil system at Eh values of 300 mv or less. An Eh value of 300 mv (at pH 7) can be assumed to be the breakpoint between oxidized and reduced zones (Figure 2).

The oxidized soil layer is characterized by a reddish brown color formed as a result of Fe^{2+} oxidation to Fe^{3+}. The grey color of the underlying reduced layer is due to Fe^{2+}. The thickness of the oxidized soil layer is determined both by O_2 concentration of the floodwater and O_2 consumption potential of the underlying soil. Howeler and Bouldin [22] have shown that the thickness of the oxidized layer is influenced by O_2 concentration in the atmosphere above the floodwater. Increased algal activity in the floodwater of a rice field can increase the O_2 concentration of the floodwater as a result of imbalance between respiration and photosynthetic activity of the algae. These conditions can in turn increase the thickness of the oxidized soil layer. Thickness of the oxidized soil layer was found to be inversely related to carbon content of the soil. As the soil O_2 demand was increased by organic matter decomposition, the thickness of the oxidized soil layer was decreased significantly. Application of inorganic fertilizers (e.g., ammonium sulfate) was found to increase the thickness of the oxidized soil layer [45].

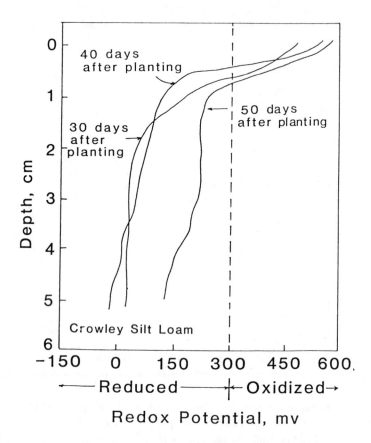

Figure 3. Thickness of oxidized soil layer of a flooded Crowley silt loam planted with rice.

Rice plants have a unique feature of transporting atmospheric O_2 through the stem to the roots, and some of this O_2 subsequently diffuses from the root into the adjacent soil layer [5, 6]. This condition creates a thin oxidized layer in the rhizosphere which can support aerobic microbial populations (Figure 1b). The development of two distinct soil layers in the rhizosphere can favor the simultaneous occurrence of nitrification in the oxidized layer and denitrification in the reduced soil layer. Thickness of the surface oxidized layer of an undisturbed soil column obtained from rice paddies was found to be less than one cm (Figure 3). At 50 days after planting, however, Eh of the soil profile had increased, indicating the oxidizing power of the rice roots.

In some rice growing areas of the world, water management practices in rice fields sometimes require draining and reflooding. This can create oxidized (drained) and reduced flooded conditions in the soil. In the rice growing areas of the United States, rice soils are maintained under flooded conditions during the rice growing period, drained at harvest, and left without flooding until the next growing season. In poorly drained soils, heavy rainfall can result in temporary flooded conditions and upon draining, oxidized conditions are restored.

Special soil and environmental conditions in a flooded rice field support two redox N processes, i.e., nitrification (oxidation of NH_4^+ to NO_3^-) and denitrification (reduction of NO_3^- to N_2). Potentially, these reactions occur in (1) continuously flooded lowland rice fields, and (2) upland rice fields which are subjected to alternate flooding and draining cycles.

Nitrification

Nitrification is a microbially mediated reaction involving the oxidation of NH_4^+ to NO_3^-. In recent years, several reviews have appeared in the literature [17, 31, 8, 57] on the microorganisms involved and the factors influencing the process in soils. To some extent, the significance of nitrification in flooded soils and sediments was also reviewed by De Datta [13], Savant and De Datta [55], and Reddy and Patrick [49]. This review will primarily focus on the role of this biochemical process in regulating forms of native soil N and applied fertilizer N in flooded soils. In flooded rice soil, nitrification can potentially occur in (1) the water column above a soil; (2) the surface oxidized soil layer; and (3) the oxidized rhizosphere of rice. Nitrification is also active in intermittently drained flooded soils.

In a flooded rice soil, the substrate for nitrification is provided from (1) ammonification (organic N to NH_4^+), and (2) application of inorganic fertilizers. In lowland rice soils, ammonification is predominantly mediated by facultative and obligate anaerobes. The characteristic features of anaerobic microbial oxidation of organic matter in lowland soils; therefore, comprise: (1) incomplete decomposition of carbohydrate into carbon dioxide, organic acids, methane, and hydrogen; (2) low energy of fermentation, resulting in the synthesis of fewer microbial cells per unit of organic carbon oxidized; and (3) low N requirements of the anaerobic metabolism.

Net release of NH_4^+ in lowland soils is determined by the ammonification and immobilization balance which is controlled by the N requirements of the microorganisms involved, nature of the organic matter, and soil and environmental factors. Agronomically, accumulation of NH_4^+ supports about 60% of the N requirements of rice. A fairly good estimate of the amount of NH_4^+ available to the rice crop can be obtained by measuring the amount of NH_4^+ accumulated during anaerobic incubation. Ammonium N accumulation in lowland rice soils was found to be rapid during the first two weeks after submergence [43]. Ammonium formed during mineralization is rapidly partitioned into (1) NH_4^+ adsorbed on the cation exchange complex, and (2) equilibrium NH_4^+ in the soil solution. Data in Figure 4 show the relative

104

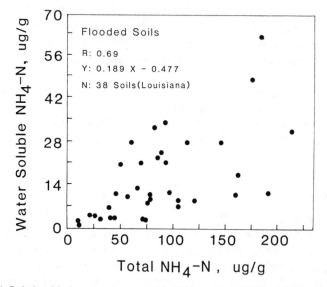

Figure 4. Relationship between water soluble NH$_4^+$ and total NH$_4^+$ of 38 flooded soils of Louisiana.

ratio of NH$_4^+$ in the soil solution to the total NH$_4^+$ present in 38 flooded soils of Louisiana. About 20% of the NH$_4^+$ was found to be in water-soluble form, while the remaining NH$_4^+$ was adsorbed on the exchange complex. Application of external sources of NH$_4^+$ can offset the equilibrium between these two fractions. The adsorptive capacity was found to be the most influential factor in the movement of NH$_4^+$ in flooded lowland soils [3] .

Ammonium present in soil solution is subjected to movement in two directions, namely (1) upward movement into the surface oxidized soil layer and floodwater, and (2) movement toward plant roots. This movement of NH$_4^+$ is accomplished by mass flow and diffusion with rate of NH$_4^+$ movement in lowland soils being governed by the concentration gradient established as a result of (1) plant uptake; (2) loss mechanisms in the rhizosphere; and (3) loss mechanisms in the surface oxidized soil layer and the floodwater. Other factors influencing the movement of NH$_4^+$ include (1) NH$_4^+$ regeneration rate in the reduced soil layer; (2) concentration of NH$_4^+$ in the pore water; (3) cation exchange capacity of the soil; (4) types of other cations on the exchange complex; and (5) relative volume of the pore space, which is a function of the bulk density.

Results in Figure 5 (a, b) show the movement of NH$_4^+$ in a lowland soil without plants. The source of NH$_4^+$ in flooded Crowley silt loam (a predominant rice soil in Louisiana) was added fertilizer NH$_4^+$ plus mineralization of native soil organic N. The source of NH$_4^+$ in flooded organic soil (lowland rice soil in south Florida) was mineralization of soil organic N. In both cases,

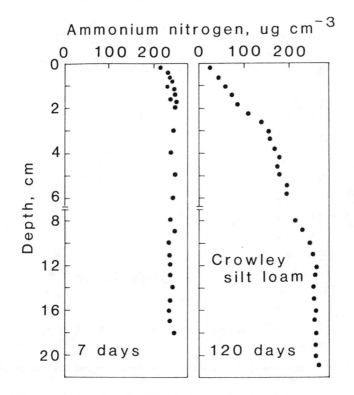

Figure 5 (a). Distribution of applied NH_4^+ in flooded Crowley silt loam with no plants (Reddy et al., 1976).

upward diffusion of NH_4^+ into the surface oxidized soil layer and the flood-water was demonstrated. Diffusion patterns of NH_4^+ will probably be different in lowland soils planted with rice and in soils where alternate flooding and draining are used as a management practice. In soils with no plants, rapid depletion of NH_4^+ in the surface soil layers was probably due to nitrification and ammonia volatilization. Results presented by Reddy and Rao [48] indicated that about 50% of the mineralized NH_4^+ was found to be lost from flooded organic soil as a result of diffusion from reduced soil to the overlying oxidized soil layer and the floodwater.

In a lowland soil planted to rice, Savant and De Datta [53, 54] and Savant et al. [56] have studied the movement of fertilizer N (prilled urea, urea super-granules, sulfur-coated urea, and urea placed in mudballs) placed in the root zone (10 cm deep). Their studies have indicated that NH_4^+ movement was downward > lateral > upward from the deep placement site. When $(NH_4)_2SO_4$ was surface applied, Bilal [9] observed appreciable concentrations of NH_4^+ in the floodwater and in the surface 1.2 cm of soil. However, when urea was surface applied, Savant and De Datta [54] measured a significant amount of NH_4^+ at the 12–14 cm depth, 4 weeks after application. Downward

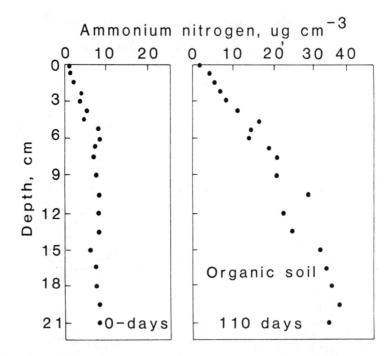

Figure 5 (b). Distribution of mineralized NH_4^+ in flooded organic soil with no plants (Reddy and Rao, 1983).

movement of plant available N can result in significant loss from coarse textured soils. Vlek et al. [69] measured serious loss in fertilizer N after placement of urea supergranules in lowland soils having a percolation rate of 5 mm day^{-1}.

Movement of NH_4^+ either upward or downward as a result of concentration gradients can decrease the amount of plant available N in the root zone, thus decreasing the fertilizer N use efficiency by rice. The transport of NH_4^+ by ionic diffusion from the reduced soil layer to the oxidized soil layer is influenced by organic matter status of the soil, presence of reduced Fe and Mn, bulk density, and rate of nitrification in the surface oxidized soil layer and in the oxidized rhizosphere.

Ammonium N diffusing into the floodwater and into the oxidized soil layer is highly unstable because of rapid oxidation to NO_3^-. Since the process of nitrification depends on the metabolism of nitrifying organisms, it is imperative that the organisms be present in adequate numbers to achieve rapid oxidation of NH_4^+. Generally, fertilized soils have larger populations of nitrifiers compared to unfertilized soils [4], and surface application of NH_4^+ fertilizers in lowland soils can potentially enhance the activity of nitrifying organisms. Nitrification rate was also found to be rapid in water containing dissolved CO_2, compared to CO_2-free water [47]. Under specialized conditions, a portion of NH_4^+ can be converted into aqueous NH_3 when pH of the floodwater is > 8.0. In the floodwater of a lowland soil, high pH conditions can exist when photosynthetic activity of algae is actively withdrawing the dissolved CO_2 from the water, thus increasing the pH at mid-day and decreasing pH at night when respiratory activities liberate free CO_2 into the water [28]. Aqueous NH_3 formation during photosynthetic periods of algae can decrease the activity of nitrifying bacteria. However, research has shown that most of the active nitrification occurs in the oxidized surface soil layer [10].

Nitrification of NH_4^+ in the floodwater and in the oxidized soil layer was observed by several research workers [11, 66, 64, 36, 9, 73, 50, 48]. Significant concentration of NO_3^- was observed in the oxidized soil layer of flooded Crowley silt loam (Figure 6). In soils with low organic matter content, the oxidized soil layer is usually thick, and most of the nitrification occurs in this zone such that no NH_4^+ diffuses into the floodwater. In soils with high organic matter content, the oxidized layer is thin, and under

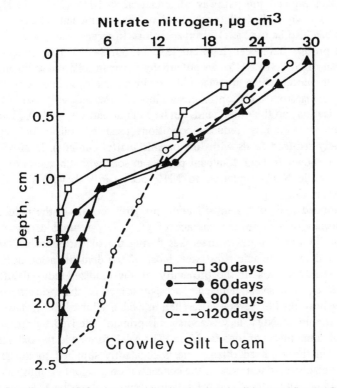

Figure 6. Distribution of NO_3^- in the oxidized and reduced soil layers of flooded Crowley silt loam soil (Reddy et al., 1976).

these conditions, NH_4^+ tends to diffuse into the floodwater. Nitrification is evidenced by the significant concentration of NO_3^- formed in the floodwater of an organic soil [48]. In a field study in California, surface application of NH_4^+ fertilizer significantly increased the NO_3^- levels in the floodwater, indicating rapid nitrification [9].

Another potential active site of nitrification in lowland rice soil is the rhizosphere. The diffusive flux of O_2 from the rice roots to the adjacent soil can create an oxidized environment around the roots. This oxidized soil rhizosphere can potentially support nitrification of NH_4^+ diffused from surrounding reduced zones. However, no experimental evidence has been reported on the significance of this process in the rice rhizosphere.

Denitrification

Under O_2-free conditions, many microorganisms can utilize NO_3^- as a terminal electron acceptor, a process called NO_3^- respiration or dissimilatory NO_3^- reduction. The pathway involving the reduction of NO_3^- to gaseous end products $(NO_3^- \rightarrow NO_2^- \rightarrow HNO \rightarrow NO \rightarrow N_2O \rightarrow N_2)$ is usually known as denitrification, and the pathway of reduction to NH_4^+ $(NO_3^- \rightarrow NH_2OH \rightarrow NH_4^+)$ is known as dissimilatory reduction. The intermediate N oxides of this process can also be used as electron acceptors. In recent years, several reviews on this process have been reported [39, 14, 63, 24, 40, 16]. Dissimilatory reduction of NO_3^- to NH_4^+ can potentially occur in soils and sediments high in organic matter which are anaerobic for long periods, including sediments, anaerobic digestors and continuously flooded lowland soils [23, 25, 62]. Denitrification, on the other hand, can be a major pathway of NO_3^- loss from soils where temporary reduced conditions exist or soils which are less intensively reduced (soils with low organic matter content). Denitrification has been shown to be a dominant process in soils with Eh values of 200 to 300 mv, while NO_3^- reduction to NH_4^+ occurs in soils with Eh values of < -100 mv [32, 12].

In flooded rice soils, denitrification primarily occurs in the reduced soil layer devoid of O_2. In the absence of O_2, NO_3^- is used as an electron acceptor by the facultative anaerobes during the oxidation of soil organic matter and other organic substrates. Most of the denitrification in lowland soils probably occurs in the proximity of the oxidized-reduced interface, which is less intensively reduced. The significance of this process has been extensively studied by several researchers as evidenced in recent reviews.

The supply of NO_3^- in lowland soils is primarily derived by nitrification of added ammoniacal or urea fertilizers. The NO_3^- formed in the oxidized soil layer or in the oxidized rhizosphere is constantly supplied to the reduced layer by diffusion in response to the concentration gradient established across the interface. The rate of denitrification is dependent on the NO_3^- supplying capacity of the system, presence of available carbon substrate, and other soil and environmental variables. On the other hand, the flux of NO_3^- is governed by the denitrification rate in the reduced soil layer, thickness of the oxidized soil layer, floodwater depth, and NO_3^- concentration in the floodwater and in the oxidized soil layer. Loss of NO_3^- from the floodwater has been found to increase with increased availability of available carbon in the underlying anaerobic soil layer [15]. Nitrate diffusion rate in flooded lowland soils has been found to be in the range of 1.2 to 1.33 cm^2 day^{-1}, which is about 7 times faster than NH_4^+ diffusion. The high diffusion coefficient values for NO_3^- are expected, since NO_3^- is an anion, is not adsorbed on the exchange complex, and tends to move rapidly in the soil water.

Nitrogen loss due to denitrification in the presence of rice plants was also found to be significant; however, the magnitude of N loss was lower compared to that for systems without plants [46]. Garcia [19, 20] has demonstrated the positive effect of the rice rhizosphere on denitrification by measuring N_2O reductase activity. Woldendorp [72] suggested that plant roots can accelerate denitrification in the rhizosphere by taking up O_2 and excreting organic substances which can serve as an energy source. Denitrifying activity in the rice rhizosphere was found to be maximum in the early stages of rice growth [21] and to decrease progressively with the age of the plant. Mandal and Datta [27] observed decreased N_2 production during denitrification as the rice plant approached maturity. Smith and Delaune [60] measured significantly greater $N_2O + N_2$ production in the first 2 days after fertilization for lowland soil planted with rice as compared to nonplanted soil. In a lowland rice soil, NO_3^- may be limiting overall N losses due to nitrification-denitrification in the rhizosphere. The controlling factor of N loss from the rhizosphere is the competition for NO_3 uptake between denitrifying bacteria and rice roots.

Nitrification-denitrification

Both nitrification and denitrification reactions are known to occur simultaneously in lowland rice fields where both oxidized and reduced zones exist. In flooded lowland soil these reactions potentially occur in (1) the surface oxidized soil layer and the underlying anaerobic soil layer, and (2) the rhizosphere. Nitrification-denitrification reactions can also occur in soils subjected to alternate flooding and draining conditions. Ammonium accumulates during flooding; then when soil is drained NH_4^+ is nitrified and NO_3^- accumulates. Upon reflooding NO_3^- can be potentially lost through denitrification.

The significance of nitrification-denitrification reactions in flooded lowland soils had been recognized as early as 1935 [58]. Later this process was confirmed by several incubation experiments [41, 59, 2, 33, 11, 35, 64, 73, 44, 50, 37]. The extent of N losses through nitrification-denitrification reactions in flooded lowland soils is dependent on the supply of NH_4^+ to the oxidized zones of the soil profile where nitrification potentially occurs, and supply of NO_3^- to reduced zones of the soil where NO_3^- reduction is potentially active. An NH_4^+ concentration gradient is established across the oxidized-reduced soil interface as a result of nitrification, while NO_3^- concentration gradient is established as a result of NO_3^- reduction in the reduced zones. Both NH_4^+ and NO_3^- diffuse into the respective zones in response to the concentration gradient. The sequential processes involved in N loss from flooded lowland soil are ammonification, NH_4^+ diffusion, nitrification, NO_3^- diffusion, and denitrification. The ultimate conversion of NH_4^+ to N_2 gas by these processes was demonstrated in flooded Crowley silt loam soil using ^{15}N [37]. From the data reported in the literature, it can be concluded that NH_4^+ diffusion and nitrification appear to be limiting N loss from most flooded lowland soils, and that NO_3^- diffusion and denitrification usually occur at a rapid rate and are not likely to limit the overall process. Nitrification reactions both in the surface (oxidized) soil layer and in the rhizosphere are controlled by O_2 diffusion and the rate of O_2 consumption by nitrifying bacteria and heterotrophic bacteria involved in organic matter decomposition.

Although controlled experiments conducted in the laboratory demonstrate that gaseous losses of N_2 and N_2O occur as a result of nitrification-denitrification, there is no direct evidence available for field conditions. However, field studies have provided indirect evidence on the possibility of these processes occurring in lowland rice soils. Results of pot and field experiments (Table 1) using ^{15}N indicate that about 35% of added N (ammonium or urea) is recovered by the plant and about 26% of the added N remains unaccounted for. Recovery of N in the greenhouse experiments was found to be generally much higher than for field experiments. Loss of N in these systems was attributed to NH_3 volatilization, nitrification-denitrification, and leaching. However, few field studies strongly suggest the possibility of nitrification-denitrification as the loss mechanism functioning in a rice field. In a field study by Patrick and Reddy [38], about 30% of the added N was lost when $^{15}NH_4$ was applied at the 7.5 cm depth (Figure 7). Serious losses were observed 4 weeks after application of fertilizer, indicating the possibility of nitrification in the rhizosphere and in the surface oxidized zone, and subsequent denitrification in the reduced zone. In another study by Wetselaar [70], about 59% of the added N was unaccounted for from surface-applied fertilizer, and 45% of the added N was not accounted for in plots receiving fertilizer by deep placement. Ammonia volatilization accounted for only 4.3% of the added N when fertilizer was surface-applied, and for only 0.7% of the added N when fertilizer N was applied by deep placement. The unaccounted-for N showed a gradual increase with time, indicating the strong possibility of coupled nitrification and denitrification. When urea was incorporated into a flooded lowland soil under tropical field conditions [26], about 31% of the added N was not accounted for, while about 10% of the added N was lost through NH_3 volatilization and about 8% was lost through

Table 1. Fate of applied ammonium sulfate and urea fertilizers in flooded soils planted with rice (greenhouse and field studies)

	Added ^{15}N fertilizer (%)	Number of studies
Plant uptake	35.1 ± 16.5	40
Loss	26.7 ± 14.0	36
Soil & Roots	38.2	

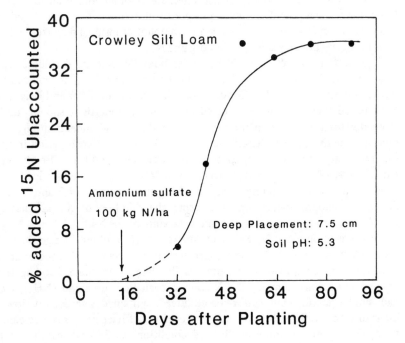

Figure 7. Percent of added [15]NH$_4$ fertilizer unaccounted for at various times during rice growing season (Patrick and Reddy, 1976b).

leaching. These results also suggest the possibility of nitrification-denitrific-ation. Freney et al. [18] and Smith et al. [61] attempted to measure N$_2$O emission under field conditions to demonstrate direct evidence for nitrifi-cation-denitrification. Nitrous oxide emissions amounted to less than 0.1% of the applied (NH$_4$)$_2$SO$_4$ or urea-N. Because of great demands for electron acceptors in the reduced zones of lowland soils, N$_2$O produced during de-nitrification is rapidly converted to N$_2$ during microbial respiration. Since N$_2$O is only a minor end product of this process, the low rates of evolution do not represent the total loss of N due to denitrification.

Water management in lowland fields can also influence the extent of N losses as a result of nitrification-denitrification. In many rice growing areas of the world, continuous flooding is usually not maintained, probably due to a shortage of water. In the United States, rice is either water-seeded or drilled-seeded. When rice is directly water seeded, soils are usually drained subsequently to lower water depth in order to enhance seedlings and root development, whereas direct-seeded fields are only flooded after the seedlings are about 10–15 cm high. In either case, alternating flooded (reduced or anaerobic) and drained (oxidized or aerobic) conditions are created. In the tropics and the subtropic areas of the world, soils used for rainfed-lowland rice cultivation may also experience alternate flooding and drained cycles, often due to heavy rainfall events and poor soil drainage. Frequent flooding and draining cycles not only decrease the amount of residual fertilizer N, but also drastically affect reserves of soil organic N.

Organic N is converted to NH_4^+ under both flooded and drained conditions, but at different rates. Under drained conditions, NH_4^+ is oxidized to NO_3^-, and when the soil is reflooded, NO_3^- is denitrified. Wijler and Delwiche [71] noted that alternating oxidized and reduced conditions should result in greater total N loss from the soil than would be found under continuously reduced conditions. The length of each oxidized and reduced period in the soil may vary, depending upon soil/environmental conditions. Reddy and Patrick [44] observed a total N loss of up to 24% from Crowley silt loam soil which underwent alternating oxidized and reduced periods of 2 and 2 days during a 4-month incubation. Increasing the duration of the alternating aerobic and anaerobic periods decreased the amounts of N loss.

In a greenhouse study, Sah and Mikkelsen [51] observed significantly lower plant available N in a soil which underwent alternate flooded-drained conditions, as compared to continuous flooding or alternate flooding and drained conditions. Among the N sources evaluated, urea and $(NH_4)_2SO_4$ had similar effects, but sulfur-coated urea maintained significantly lower levels of plant available N in the soil. Nitrogen use efficiency by rice decreased from 67% under continuously flooded conditions to 35% when soil underwent two cycles of alternate flooded and drained conditions [52]. Nitrification inhibition can significantly prevent the loss of added and native N in rice soils managed under alternate flooded and drained conditions.

Conclusions

Extensive laboratory incubation studies reported in the literature clearly establish the role of nitrification-denitrification reactions in N loss from flooded lowland soils. Despite many greenhouse and field experiments conducted in serveral parts of the world to evaluate the fate of applied fertilizer ^{15}N in lowland soil, quantitative information identifying the key processes involved is still not available. None of the field studies using ^{15}N have documented direct evidence (measuring gaseous end products) of nitrification-denitrification. Mass balance of ^{15}N in the field studies attributes the unaccounted-for N to potential loss via nitrification-denitrification, NH_3 volatilization, and leaching.

Nitrogen loss due to nitrification-denitrification can potentially occur when fertilizer is either surface applied or placed in the root zone. Surface-applied fertilizer, if in ammoniacal form, can be rapidly converted to NO_3^-, since the applied fertilizer granules sink through the floodwater and settle on the surface oxidized layer. Slow-release fertilizers can reduce losses due to this process. However, when urea fertilizer is surface applied, significant amounts of N can be lost through volatilization, since hydrolysis of urea alters the alkalinity of the floodwater [67, 68]. Although NH_3 volatilization losses can be reduced by placing the fertilizer in the root zone [28], it is still unknown whether N loss due to nitrification-denitrification can be prevented. An active (aerobic) rhizosphere can potentially increase the rate of nitrification and subsequent denitrification, thus decreasing the amount of plant available N in the root zone. Future research in this area is urgently needed in order to design better management practices to increase the efficiency of N utilization by rice and to conserve the amount of fertilizer to be applied.

Nitrogen losses due to nitrification-denitrification reactions can be prevented if (1) nitrification is prevented, thus maintaining inorganic N in NH_4^+ form; (2) placing the fertilizer in the root zone; and (3) increasing the O_2 demand in the root zone by increasing the organic matter content of the soil. These are some of the potential management strategies that can be used to prevent N losses from rice fields.

References

1. Alberda T (1953) Growth and root development of lowland rice and its relation to oxygen supply. Plant and soil 5:1–28
2. Amer F (1960) Evaluation of dry sub-surface and wet surface application for rice. Plant and Soil 13:47–54
3. Aomine S (1978) Movement of ammonium in paddy soils in Taiwan. Soil Sci Plant Nutr 24:571–580
4. Ardakani MS, Schulz RK and McLaren AD (1974) A kinetic study of ammonium and nitrite oxidation in a soil field plot. Soil Sci Soc Am Proc 38:273–277
5. Armstrong W (1964) Oxygen diffusion from the roots of some British bog plants. Nature 204(4960):801–802
6. Armstrong W (1967) The relationship between oxidation-reduction potentials and oxygen-diffusion levels in some waterlogged organic soils. J Soil Sci 18(1):27–34
7. Armstrong W and Boatman DJ (1967) Some field observations relating the growth of bog plants to conditions of soil aeration. J Ecol 55:101–110
8. Belser LW (1979) Population ecology of nitrifying bacteria. Ann Res Microbiol 33:309–333
9. Bilal I (1974) Transformations and transport of applied NH_4-N in flooded rice culture. Ph D dissertation, Univ California, Davis, 139 pp, Xerox Univ Microfilms, 75–831b
10. Billen G (1975) Nitrification in the Scheldt Estuary (Belgium and the Netherlands). Estuarine and Coastal Marine Science 3:79–89
11. Broadbent FE and Tusneem ME (1971) Losses of nitrogen from some flooded soils in tracer experiments. Soil Sci Soc Am Proc 35:922–926

12. Buresh RJ and Patrick, Jr., WH (1981) Nitrate reduction to ammonium and organic nitrogen in an estuarine sediment. Soil Biol Biochem 13:279–283
13. De Datta SK (1981) Principles and practices of rice production. John Wiley & Sons, New York, 618 p
14. Delwiche CC and Bryan BA (1976) Denitrification. Ann Rev Microbiol 30:241–262
15. Engler RM and Patrick, Jr., WH (1974) Nitrate removal from floodwater overlying flooded soils and sediments. J Environ Qual 3:409–413
16. Firestone MK (1982) Biological denitrification. In Nitrogen in Agricultural Soils, pp 289–326. Madison, Wisc. Am Soc of Agron
17. Focht DD and Verstraete W (1977) Biochemical ecology of nitrification-denitrification. Adv Microbiol Ecol 1:124–204
18. Freney JR, Denmead OT, Watanabe I and Craswell ET (1981) Ammonia and nitrous oxide losses following applications of ammonium sulfate to flooded rice. Aust J Agric Res 32:3745
19. Garcia JL (1975a) Effect rhizosphere du riz sur la denitrification. Soil Biol Biochem 7:139–141
20. Garcia JL (1975b) Evaluation de la denitrification dans les rizieres par la methode de reduction de N_2O. Soil Biol Biochem 7:251–256
21. Garcia JL (1977) Cah OSTROM. Ser Biol 12:83–87 (cited by Savant and De Datta, 1982)
22. Howeler RH and Bouldin DR (1971) The diffusion and consumption of oxygen in submerged soils. Soil Sci Soc Am Proc 35:202–208
23. Keeney DR, Chen RL and Graetz DA (1971) Importance of denitrification and nitrate reduction in sediments to the nitrogen budget of lakes. Nature 233:66–67
24. Knowles R (1981) Denitrification. In Clark FE and Rosswall T, eds. Terrestrial nitrogen cycles, 33:315–329, Ecol Bull (Stockholm)
25. Koike I and Hattori A (1978) Denitrification and ammonia formation in anaerobic coastal sediments. Appl Environ Microl 35:278–282
26. Krishnappa AM and Shinde JE (1978) Working paper No. 41. Presented at the 4th Res coordination meeting of the joint FAO/1AEA/GSF coordinated program on N residues, Piracicaba, Brazil (cited by Savant and De Datta, 1982)
27. Mandal SR and Datta NP (1975) Direct quantitative estimation of nitrogen gas loss under submerged rice crops – a study with [14]nitrogen. Indian Agric 19:127–134
28. Mikkelsen DS, De Datta SK and Obcemea WN (1978) Ammonia volatilization losses from flooded rice soils. Soil Sci Soc Am J 42:725–730
29. Mortimer CH (1941) The exchange of dissolved substances between water and mud in lakes. J Ecol 29:280–329
30. Mortimer CH (1971) Chemical exchanges between sediments and water in the great lakes – speculations on probable regulatory mechanisms. Limnol Oceanogr 16:387–404
31. Painter HA (1970) A review of literature on inorganic nitrogen metabolism in microorganisms. Water Research 4:393–450
32. Patrick Jr WH (1960) Nitrate reduction rates in a submerged soil as affected by redox potential. 7th Intern Congress of Soil Sci Madison, Wisc. 2:494–500
33. Patrick Jr WH and Wyatt R (1964) Soil nitrogen loss as a result of alternate submergence and drying. Soil Sci Soc Am Proc 28:647–653
34. Patrick Jr WH and Delaune RD (1972) Characterization of the oxidized and reduced zones in flooded soil. Soil Sci Soc of Am Proc 36:573–576
35. Patrick Jr WH and Tusneem ME (1972) Nitrogen loss from flooded soil. Ecol 53:735–737
36. Patrick Jr WH and Gotoh S (1974) The role of oxygen in nitrogen loss from flooded soils. Soil Sci 118:78–81
37. Patrick Jr WH and Reddy KR (1976a) Nitrification-denitrification in flooded soils and sediments: Dependence on oxygen supply and ammonium diffusion. J Environ Qual 5:469–472
38. Patrick Jr WH and Reddy KR (1976b) Fate of fertilizer nitrogen in flooded soil. Soil Sci Soc Am J 40:678–681

39. Payne WJ (1973) Reduction of nitrogenous oxides by microorganisms. Bacteriol Rev 37:409–452
40. Payne WJ (1981) Denitrification. John Wiley & Sons, New York, NY p. 214
41. Pearsall WH (1950) The investigation of wet soils and its agricultural implications. Emp J Agr 18:289–298
42. Pearsall WH and Mortimer CH (1939) Oxidation-reduction potentials in water-logged soils, natural waters, and muds. J Ecol 27:483–501
43. Ponnamperuma FN (1972) The chemistry of submerged soils. Advan Agron 24:29–96
44. Reddy KR and Patrick Jr WH (1975) Effect of alternate aerobic and anaerobic conditions on redox potential, organic matter decomposition, and nitrogen loss in a flooded soil. Soil Bio Biochem 7:87–94
45. Reddy KR and Patrick Jr WH (1977) Effect of placement and concentration of applied $^{15}NH_4$-N on nitrogen loss from flooded soil. Soil Sci 123:142–147
46. Reddy KR and Patrick Jr WH (1980) Losses of applied $^{15}NH_4$-N, urea, ^{15}N, and organic ^{15}N in flooded soil. Soil Sci 130:326–330
47. Reddy KR and Graetz DA (1981) Use of shallow reservoirs and flooded organic soil systems for wastewater treatment: Nitrogen and phosphorus removal. J Environ Qual 10:113–119
48. Reddy KR and Rao PSC (1983) Nitrogen and phosphorus fluxes from flooded organic soil. Soil Sci 136:300–307
49. Reddy KR and Patrick Jr WH (1984) Nitrogen transformations and loss in flooded soils and sediments. CRC Critical Reviews in Environ Control 13:273–309
50. Reddy KR, Patrick Jr WH and Phillips RE (1976) Ammonium diffusion as a factor in nitrogen loss from flooded soils. Soil Sci Soc Am J 40:528–533
51. Sah RN and Mikkelsen DS (1983a) Availability and utilization of fertilizer nitrogen by rice under alternate flooding. I. Kinetics of available nitrogen under rice culture. Plant Soil 75:221–226
52. Sah RN and Mikkelsen DS (1983b) Availability and utilization of fertilizer nitrogen by rice under alternate flooding. II. Effects on growth and nitrogen use efficiency. Plant Soil 75:227–234
53. Savant NK and De Datta SK (1979) Nitrogen release patterns from deep placement sites of urea in a wetland rice soils. Soil Sci Soc Am J Proc 43:131–134
54. Savant NK and De Datta SK (1980) Movement and distribution of ammonium N following deep placement of urea in wetland rice soil. Soil Sci Soc Am J 44:559–565
55. Savant NK and De Datta SK (1982) Nitrogen transformations in wetland rice soils. Adv Agron 35:241–302
56. Savant NK, De Datta SK and Craswell ET (1982) Distribution patterns of ammonium nitrogen and ^{15}N uptake by rice after deep placement of urea supergranules in wetland soil. Soil Sci Soc Am J 46:567–573
57. Schmidt EL (1982) Nitrification in soil. In Nitrogen in agricultural soils, pp 253–288, Madison, Wisc., American Society of Agronomy
58. Shioiri M and Mitsui S (1935) J Sci Soil and Manure, Japan 9:261–268 quoted by Mitsui, 1954
59. Shioiri M and Tanada T (1954) The chemistry of paddy soils in Japan. Ministry of Agriculture and Forestry. Tokyo
60. Smith CJ and DeLaune RD (1984) Effect of rice plants on nitrification-denitrification loss of nitrogen. Plant Soil 79:287–290
61. Smith CJ, Brandon M and Patrick Jr WH (1982) Nitrous oxide emission following urea-N fertilization of wetland rice. Soil Sci Plant Nutr 28:161–171
62. Sorensen J (1978) Capacity for denitrification and reduction of nitrate to ammonia in coastal sediment. Appl Environ Microbiol 35:301–305
63. Stouthamer AH, van't Riet J and Oltmann LF (1980) Respiration with nitrate as acceptor. In Knowles CJ, ed. Diversity of bacterial respiratory systems Vol 2, pp 19–48, Boca Raton, Fla, CRC Press Inc
64. Takai Y and Uehara Y (1973) Nitrification and denitrification in the surface layer of submerged soil. J Sci of Soil and Manure 44:463–502
65. Turner FT and Patrick Jr WH (1968) Chemical changes in waterlogged soils as a result of oxygen depletion. Trans 9th Inter Congr Soil Sci 4:53–65

66. Tusneem ME and Patrick, JR, WH (1971) Nitrogen transformations in waterlogged soil. Agric Exp Sta, LA State Univ, Bull 657
67. Vlek PLG and Stumpe JM (1978) Effects of solution chemistry and environmental conditions on ammonia volatilization losses from aqueous systems. Soil Sci Soc Am J 42:416–421
68. Vlek PLG and Craswell ET (1979) Effect of nitrogen source and management on ammonia volatilization losses from flooded rice soil systems. Soil Sci Soc Am J 43:352–358
69. Vlek PLG, Brynes BH and Craswell ET (1980) Effect of urea placement on leaching losses of nitrogen from flooded rice soils. Plant Soil 54:441–449
70. Wetselaar R (1975) Proc Conf Thai-Aust Chao phya Res Proj Chainat. 1966–1975 pp 91–100 (cited by Savant and De Datta 1982 Adv Agron 35:241–302)
71. Wijler J and Delwiche CC (1954) Investigation on the denitrifying process in soil. Plant and Soil 5:155–169
72. Woldendorp JM (1963) The influence of living plants on denitrification. Meded Landbouwhogeschool, Wageningen 63:1–100
73. Yoshida T and Padre BC (1974) Nitrification and denitrification in submerged Maahas clay soil. Soil Sci Plant Nutr 20:241–247

6. Ammonium dynamics of puddled soils in relation to growth and yield of lowland rice

K MENGEL,[1] HG SCHON,[1] G KEERTHISINGHE[2] and SK DE DATTA

International Rice Research Institute, P.O. Box 933, Manila, Philippines

Key words: exchangeable NH_4^+, nonexchangeable NH_4^+, ^{15}N tracer technique, NH_4^+ fixation, vermiculite

Abstract. The release of non-exchangeable (fixed) NH_4^+ and the importance of exchangeable NH_4^+ at transplanting (initial exchangeable NH_4^+) for rice (*Oryza sativa* L.) growth was studied in representative lowland rice soils of the Philippines.

The experiments showed that initial exchangeable ammonium behaved like fertilizer N and thus may serve as a valuable guideline for nitrogen fertilizer application rates when calculated on a hectare basis. By using the ^{15}N tracer technique it was found that non-exchangeable ammonium in soil may contribute to the nitrogen supplying capacity of lowland rice soils. Fixation and release of NH_4^+ seem to be more dependent on the form of clay minerals than on clay content. In soils rich in vermiculite non-exchangeable ammonium should be considered together with other available N sources such as exchangeable ammonium for N fertilizer recommendations for lowland rice.

Introduction

Ammonium nitrogen is the dominant form of mineral nitrogen in lowland rice soils, existing in three major fractions:
— Ammonium in soil solution
— Ammonium at exchange sites
— Ammonium in non-exchangeable form
Ammonium nitrogen in soil solution and at exchange sites is readily available to the rice plant, whereas reports on the availability of non-exchangeable ammonium are conflicting. Walsh and Murdock [25] and Martin et al. [13] observed a very low availability of non-exchangeable ammonium to upland crops. In contrast to these studies, much higher release of this fraction to upland crops was reported by Mohammed [15], Kowalenko and Ross [12] and Mengel and Scherer [14]. Most of the experiments concerning this problem were conducted on upland soils [16]. Broadbent and Nakashima [1] and Broadbent and Tusneem [2] showed that under flooded conditions substantial amounts of NH_4^+ could be fixed by Sacramento clay. These

[1] Present address: Justus Liebig Universitat Giessen, D-6300 Giessen, Sudanlage 6, West Germany and [2] Present address: No. 10 Mahamaya Mawatha, Kandy, Sri Lanka

117

authors, however, did not investigate whether the fixed NH_4^+ could be subsequently released to plants. Using ^{15}N tracer technique, the availability of non-exchangeable ammonium in lowland rice soils was determined [10] under field conditions.

Investigations in recent years support the concept that initial mineral N content (ammonium N + nitrate N) can be used to predict N fertilizer needs of upland crops [11, 26, 21, 23, 4, 24]. Based on these results, we conducted field experiments to study whether the amount of exchangeable ammonium at the beginning of a planting season (initial exchangeable ammonium) is correlated to N-uptake and grain yield of rice. We did not include any other N-forms such as nitrate, assuming that ammonium is the most important N-form for lowland rice.

Materials and methods

Release of non-exchangeable ammonium (^{15}N labelled) in lowland rice soils

During the 1981 wet season, soil samples were taken from the ongoing long-term nitrogen response experiments conducted by the Agronomy Department (IRRI) at the Maligaya, Cabuyao and Iloilo sites. Soils were sampled from off the top 20 cm layer. The important soil characteristics are shown in Table 1. From the soil samples of each experimental site, wet soil was taken equivalent to 2.5 kg of oven dry soil. ^{15}N (96.5 N atom %) in form of $^{15}NH_4^+$ Cl (0.9 g) was dissolved in 300 ml of deionized water and added to each of these samples. The total amount of water in the samples was adjusted to 3 l considering the moisture content of the soil to obtain a soil:solution ratio of 2.5:3 and a ^{15}N concentration of 300 ppm.

After adding water, the samples were agitated several times and incubated at 40°C for 1 week. During incubation, the suspensions were mixed thoroughly. After incubation the samples were transferred to columns and the NH_4^+ was completely exchanged by 1 N $CaCl_2$ and eluted by washing the

Table 1. Soil characteristics of the three experimental sites

	Maligaya	Cabuyao	Iloilo
Soil texture	Silty clay loam	Clay	Clay
pH (1:1 H_2O)	5.8	6.8	6.4
Organic C (%)	1.42	1.83	1.76
Total N (%)	0.13	0.22	0.15
CEC (meq/100 g)	32	45	52
Nonexchangeable NH_4^+-N			
in ppm	56	59	194
in % total N	4	3	13
Clay mineralogy	Montmorillonite Vermiculite (major)	Montmorillonite Vermiculite (minor)	Montmorillonite Practically all
Soil series	Maligaya	Guadalupe	Sta. Rita
Soil suborders	Typic Pellusterts	Typic Pellusterts	Typic Pelluderts

soil columns several times with deionized water. Then the soil was taken out of the columns bulked together and dried at $35°C$. After drying, the samples were crushed and passed through a 2-mm sieve. The sieved soil was placed into soil holders as described by Mengel and Scherer [14]. The soil holders were PVC rings, 10 cm in diameter, 2.8 cm high, each side being covered by stainless steel net with mesh size 0.025 mm allowing only solute flux in and out. The holders were filled with 200 g of ^{15}N-labelled soil. Nine such holders were placed on each of the three experiment sites (Maligaya, Cabuyao and Iloilo). The soil holders were placed horizontally at 20 cm depth at the time of transplanting and remained in the soil until harvest. Three nitrogen fertilizer levels per site were selected from the N-response experiments. One holder per nitrogen treatment was placed in the soil in field experimental sites. Each nitrogen treatment was replicated 3 times.

Analytical procedure

The soil samples were analyzed for labelled non-exchangeable ammonium and total non-exchangeable ammonium (labelled + nonlabelled NH_4^+) prior to their placement in the field according to the method of Silva and Bremner [20]. Each analysis was carried out with 4 replications. At harvest, the soil holders were dug out and the content of labelled non-exchangeable NH_4^+ and total non-exchangeable NH_4^+ was analyzed. From each soil holder, two sub-samples were analyzed for ammonium according to the Kjeldahl technique of Keeney and Bremner [9].

The tops of four rice hills surrounding each soil holder were harvested for ^{15}N and total N analyses. The ^{15}N in soil and plant samples was determined according to the technique of Faust [5], by measuring the emission spectrum of labelled N_2 with a Statron ^{15}N analyzer. Determination of exchangeable NH_4^+ of soil samples was carried out as follows.

The soil samples were homogenized and frozen directly after collection to prevent microbial activity. After thawing the samples they were immediately extracted for exchangeable NH_4^+ by 2 N KCl [3]. NH_4^+ was determined colorimetrically by using indophenol blue according to the method of Harwood and Kuhn [7]. Total N of soil and plant samples was assessed by the micro Kjeldahl technique.

Field experiments

In the 1980 wet season and the 1981 dry season initial (at transplanting) exchangeable NH_4^+ was determined in randomized complete block nitrogen response experiments in farmers' fields. All sites were irrigated.

Selected soil properties of the sites are given in Table 2. Soil samples were taken just before fertilizer application and transplanting. One sample was taken per $16 m^2$ soil surface using a PVC pipe with 5.3 cm of diameter to a depth of 15 cm.

The N fertilizer treatments are given in Table 3 and were replicated three

120

Table 2. Soil characteristics of locations investigated for exchangeable soil NH_4^+ and crop response. Initial exchangeable NH_4^+ was calculated for 15 cm soil depth assuming a bulk density of 1

1980 wet season

Soil no.	1	2	3	4	5	6	7
Location	Pila	Lamut	Bunol	Sto. Domingo	Bakal	Curba	Masit
Clay %	51	37	46	56	32	36	35
pH (1:1 H_2O)	6.9	6.2	6.7	6.9	5.3	6.3	6.6
Total N %	0.25	0.19	0.1	0.1	0.13	0.13	0.19
CEC me/100 g	34	31	34	35	23	25	29
Initial exch. NH_4^+ kg N ha^{-1}	62	27	12	16	29	37	47

1981 dry season

Soil no.	9	10	11	12	13	14	15	16	17
Location	Masit	Bunol	Balabat	Labuin	Muñoz	Sto. Domingo	Bakal	San Antonio	Laguna
Clay %	37	46	51	34	30	57	30	49	51
pH (1:1 H_2O)	6.1	6.9	6.2	6.0	5.8	6.8	5.9	5.9	5.9
Total N %	0.2	0.1	0.21	0.14	0.1	0.09	0.11	0.16	0.22
CEC me/100 g	25	24	28	21	18	30	20	27	26
Initial exch. NH_4^+ kg N ha^{-1}	22	15	37	36	16	17	33	35	48

Table 3. Description of nitrogen fertilizer treatments

Treatment no.	Form of N[a]	N-application rate[b] (kg N/ha)					Remarks
		B & I	PPL	Top	PI	Total	
1	–	–	–	–	–	–	Check
2	Urea	–	–	15[c]	14	29	Farmers' split
3	Urea	20	–	–	9	29	Researchers' split
4	SCU	29	–	–	–	29	Researchers' split
5	USG	–	29	–	–	58	Researchers' split
6	Urea	–	–	29	29	58	Farmers' split
7	Urea	39	–	–	19	58	Researchers' split
8	SCU	58	–	–	–	58	Researchers' split
9	USG	–	58	–	–	58	Researchers' split
10	Urea	–	–	44	43	87	Farmers' split
11	Urea	58	–	–	29	87	Researchers' split
12	SCU	87	–	–	–	87	Researchers' split
13	USG	–	87	–	–	87	Researchers' split
14	Urea	–	–	58	58	116	Farmers' split
15	Urea	77	–	–	39	116	Researchers' split
16	SCU	116	–	–	–	116	Researchers' split
17	USG	–	116	–	–	116	Researchers' split
18	Urea	96	–	–	49	145	Researchers' split
19	Urea	–	–	73	72	145	Farmers' split

Treatments 16–19 only in dry season

[a] Urea = prilled (46% N); SCU = sulfur-coated urea (forestry grade), 36.6% N with the release of 20.7% N in 7 days after application; USG = urea supergranule
[b] B & I = basal broadcast and incorporated before last harrowing. PPL = point placement 2–4 days after transplanting at the center between every 4 hills. Top = topdress at 10 DT. PI: researchers' practice – 5–7 days before panicle initiation; farmers' practice – past panicle initiation
[c] Topdress at 20 days after transplanting

times. Plot size was 4×4 m. Plant spacing was 20×20 cm. Test cultivar was IR42 during the 1980 wet season, whereas the IR varieties varied from site to site during the 1981 dry season. To determine plant N-uptake (only in the 1980 wet season) and grain yield, tops of plants were harvested at full maturity from the centre of each plot ($5 m^2$ soil surface).

Results

Release of non-exchangeable NH_4^+ in lowland soils

Table 4 shows the amount of total non-exchangeable ammonium (labelled + unlabelled) present at the beginning and at the end of the experiment. A highly significant decrease in non-exchangeable ammonium was observed in all locations at all N-levels with one exception (Sta. Rita clay without fertilizer treatment).

Nitrogen-fertilization had no clear effect on the release of non-exchangeable ammonium. Calculating the decrease in total non-exchangeable NH_4^+ on a

Table 4. Total non-exchangeable ammonium content of the incubated soil at the beginning and at the end of the experiment

Soil	Nitrogen applied (kg N/ha)	NH_4^+-N in oven dry soil (ppm)		
		Beginning a	End b	Δ a — b
Maligaya silty clay loam	0		97.1[+++]	− 47
	60	143.9	98.2[+++]	− 46
	120		104.6[+++]	− 39
Guadalupe clay	0		62.4[++]	− 12
	80	74.0	64.4[+++]	− 10
	120		62.7[+++]	− 11
Sta. Rita clay	0		265.8[ns]	+ 20
	60	245.7	221.1[++]	− 21
	120		220.0[++]	− 26

[+++] Significant difference at 0.1% level
[++] Significant difference at 1.0% level

hectare basis and assuming a soil bulk density of 1 and a soil layer of 20 cm, approximately 80, 20 and 40 kg N ha^{-1} were released during the growth period from the Maligaya silty clay loam, the Guadalupe clay and the Sta. Rita clay, respectively. At all locations, a release of non-exchangeable labelled NH_4^+ was found (Table 5). The amounts released were equal to 60, 16 and 40 kg N ha^{-1} in the Maligaya silty loam, the Guadalupe clay and the Sta. Rita clay, respectively, assuming a soil depth of 20 cm and a soil bulk density of 1. Although no net release of total non-exchangeable NH_4^+ was found in the fertilizer N_0 treatment of the Sta. Rita clay loam (Table 4), the same soil showed a highly significant release of labelled non-exchangeable NH_4^+ (Table 5). These data indicate that the released labelled NH_4^+ was replaced by nonlabelled NH_4^+. Obviously NH_4^+ produced by microbial decomposition

Table 5. Content of ^{15}N-labelled non-exchangeable ammonium at the beginning and at the end of the experiment

Soil	Nitrogen applied (kg N/ha)	NH_4^+-N in oven dry soil (ppm)		
		Beginning a	End b	Δ a — b
Maligaya silty clay loam	0		41.6[+++]	− 35
	60	77.0	42.4[+++]	− 34
	120		47.8[+++]	− 29
Guadalupe clay	0		12.4[+++]	− 8
	80	20.6	12.5[+++]	− 8
	120		12.2[+++]	− 8
Sta. Rita clay	0		29.8[+++]	− 21
	60	51.4	27.7[+++]	− 24
	120		29.5[+++]	− 21

[+++] Significant difference at 0.1% level

Table 6. ^{15}N content of rice plants growing near the soil holder and % proportion of ^{15}N from total N uptake of rice tops. Total N uptake $= 100\%$

Soil	Nitrogen applied (kg N/ha)	^{15}N/hill (mg)	^{15}N in relation to total N uptake (%)
Maligaya silty clay loam	0	4.1	1.8
	60	3.7	1.0
	120	4.3	0.8
Guadalupe clay	0	4.4	1.4
	80	5.0	1.4
	120	6.0	1.3
Sta. Rita clay	0	5.4	2.2
	60	5.8	2.1
	120	5.5	1.5

during the growth period was refixed by clay minerals. Although the soil in the soil holders was not directly in contact with rice roots, labelled N was taken up by the crop (Table 6).

Relations between initial NH_4^+ and N uptake and grain yield of rice

Initial NH_4^+ content in the analyzed soils ranged from 12 to 62 kg N ha^{-1} in the 1980 wet season and from 16 to 48 kg N ha^{-1} (Table 2) in the 1981 dry season.

The relations between the N fertilizer rate and the N uptake of the rice crop in the prilled urea farmers' practice treatment are shown in Figure 1. The relations were substantially improved when the N-uptake of the crop was plotted against fertilizer N + initial exchangeable soil NH_4^+ (Figure 2).

The widely differing N-uptake patterns between the experimental sites were not evident, if the initial exchangeable soil NH_4^+ was taken into consideration. What was found to be true for the farmers' practice treatment was also true for the other N-application treatments.

The coefficients of determinations (R^2) were higher when N-uptake was correlated with fertilizer N + initial exchangeable NH_4^+ as compared with the N uptake versus fertilizer N only (Table 7). The initial exchangeable NH_4^+ was also more closely correlated with N uptake than was with total soil N and N uptake (Total N: $R^2 = 0.59$, initial exchangeable NH_4^+: $R^2 = 0.76$).

Close linear relations were also obtained between the initial exchangeable NH_4^+ in the plots without fertilizer N and the relative grain yield (Figures 3, 4). Relative grain yield is defined as the % grain yield from the maximum grain yield ($= 100$) attained by N fertilizer on the corresponding location in the same season. Figure 5 shows the relations between fertilizer N rate and relative grain yield in the urea researchers' practice treatment. A great variation in grain yield among different sites was evident, particularly in the fertilizer N_0 treatments. If, however, the relative grain yield was plotted against the initial exchangeable NH_4^+ + fertilizer N, a Mitscherlich — type

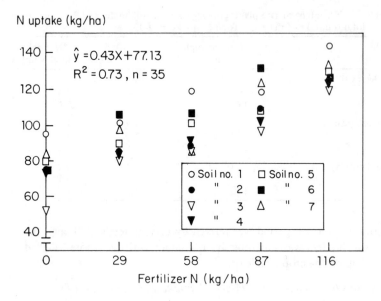

Figure 1. Correlation between fertilizer N rate and total N uptake by rice grown on 7 soils using farmers' N management practices. 1980 Wet season

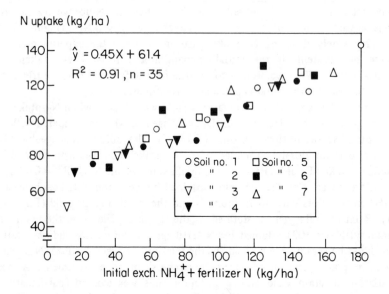

Figure 2. Correlation between initial exchangeable NH_4^+-N + fertilizer N rate and total N uptake by rice on 7 soils using farmers' N management practices. 1980 Wet season

Table 7. Coefficients of determination for the relations between (A) N fertilizer rate and N uptake and (B) N fertilizer rate + initial exchangeable NH_4^+ and N uptake. 1980 Wet season

Treatment	N fertilizer R^2 (A)	N fertilizer + NH_4^+ R^2 (B)
Prilled urea researchers' practice	0.69***	0.75***
Prilled urea farmers' practice	0.73***	0.91***
Sulfur-coated urea	0.73***	0.83***
Urea supergranules	0.66***	0.81***

*** Significant at 0.1% level

of crop response was obtained (Figure 6). The other N application treatments showed higher coefficients of determination if the relative grain yield was plotted with initial exchangeable NH_4^+ + fertilizer N as compared with the relative grain yield versus fertilizer N only (Table 8).

Total soil N on the other hand – with one exception – did not correlate with the relative grain yield (Table 9).

Discussion

The net release of non-exchangeable NH_4^+ during the growing season was

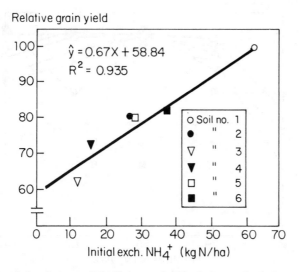

Figure 3. Correlation between NH_4^+-N in non-fertilized plots and relative grain yield of rice grown on 6 soils. 1980 Wet season

126

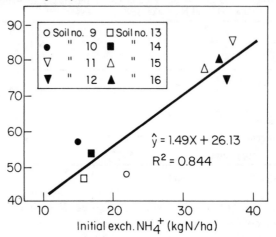

Figure 4. Relations between initial NH_4^+-N in non-fertilized plots and relative grain yield of rice grown on 8 soils. 1981 Dry season

highest in the Maligaya silty clay loam and lowest in the Guadalupe clay soil. The relatively high net release in Maligaya silty clay loam may be related to the high vermiculite content of that soil. The amounts of released labelled non-exchangeable ammonium in the Maligaya silty clay loam and in the

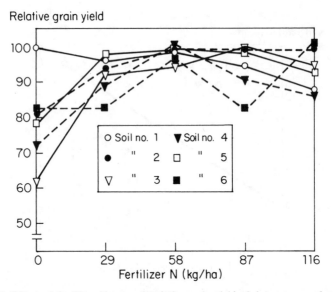

Figure 5. Effect of fertilizer N rates on relative grain yield of rice grown on 6 soils using prilled urea with researchers' split. 1980 Wet season

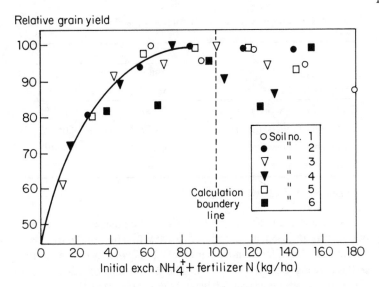

Figure 6. Correlation between initial exchangeable NH_4^+-N + fertilizer N rate and grain yield of rice grown on 6 soils using prilled urea with researchers' split. 1980 Wet season

Guadalupe clay were lower than the total release of non-exchangeable NH_4^+, indicating that besides the $^{15}NH_4^+$ a substantial amount of $^{14}NH_4^+$ was also released by the clay minerals. In the N zero treatment (without fertilizer N) of the Sta. Rita clay no significant change of total non-exchangeable NH_4^+ was found (Table 4). Yet there was a significant decrease in labelled non-exchangeable NH_4^+ (Table 5) suggesting that besides a release of NH_4^+, also a refixation of NH_4^+ had occurred.

Table 8. Coefficients of determination for the relations between (A) N fertilizer rate and relative grain yield and (B) fertilizer N + initial exchangeable NH_4^+ and relative grain yield

Treatment	N fertilizer R^2 (A)	N fertilizer + NH_4^+ R^2 (B)
Wet season 1980		
Prilled urea researchers' practice	0.54***	0.87***
Prilled urea farmers' practice	0.75***	0.81***
Sulfur-coated urea	0.67***	0.76***
Urea supergranules	0.57***	0.77***
Dry season 1981		
Prilled urea researchers' practice	0.55***	0.72***
Prilled urea farmers' practice	0.55***	0.68***
Sulfur-coated urea	0.57***	0.75***
Urea supergranules	0.51***	0.83***

***Significant at 0.1% level

Table 9. Correlation coefficients for total soil N and relative grain yield of rice

| | R^2 | |
| | 1980 | 1981 |
Fertilizer treatment	wet season	dry season
Prilled urea researchers' practice	0.757^+	0.039^{ns}
Prilled urea farmers' practice	0.412^{ns}	0.057^{ns}
Sulfur-coated urea	0.408^{ns}	0.068^{ns}
Urea supergranules	0.466^{ns}	0.036^{ns}

$^+$ Significant at 5% level

In all soils analyzed, about 40 to 50% of the labelled non-exchangeable ammonium was released during the crop growth. The uptake of ^{15}N by the rice plants revealed that non-exchangeable NH_4^+ was available to the crop during its growing period (Table 6). The proportion of labelled N in the crop decreased with increasing fertilizer N indicating that the labelled N was diluted by fertilizer N. If the $^{15}NH_4^+$ in the soil holder would have been available to the same extent as the N outside the soil holder, the % proportion of ^{15}N from the total plant N should have amounted to 2.5%. The actual figures for the % proportion are somewhat lower (Table 6). Considering that the roots were not in direct contact with the ^{15}N labelled soil, the availability of non-exchangeable labelled NH_4^+ was high.

Release of non-exchangeable NH_4^+ was particularly high in the Maligaya soil amounting to about $80\,kg\,N\,ha^{-1}$ if a soil depth of 20 cm and a soil bulk density of 1 is assumed. The Maligaya silty clay loam, although the lowest in clay content from the three soils studied, was rich in vermiculite. Therefore, soils rich in vermiculite may release higher amounts of non-exchangeable NH_4^+ and also refix NH_4^+ if NH_4^+-containing fertilizer is applied [10]. Vermiculite is known to release easily non-exchangeable K^+ [19]. Similar processes obviously occur for NH_4^+. According to Sippola et al. [22] vermiculite fixes NH_4^+ preferentially to K^+.

Total N of flooded rice soils seems to be a poor indicator for the N availability (Table 9). These findings are in good agreement with other authors [8, 6].

On all soils investigated, the exchangeable NH_4^+ had a clear impact on the N uptake and grain yield of rice (Figures 2, 3, 4, 6, Tables 7, 8). The quantities of exchangeable NH_4^+ found at transplanting were relatively low (Table 2) as compared with the total N in the top soil layer of flooded rice soils [17]. Nevertheless the exchangeable soil NH_4^+ seems to be of major importance for the N supply of the rice crop. The significant correlations between initial exchangeable NH_4^+ + fertilizer N versus N uptake and grain yield of the crop reveal that the exchangeable NH_4^+ behaves like fertilizer N and thus should be taken into consideration when assessing N fertilizer rates. For example, in soils with a high level of exchangeable NH_4^+ low N rates should be applied and vice versa [18]. Based on the amount of initial exchangeable NH_4^+ (kg N ha^{-1} in the top layer) a soil test and N fertilizer recommendation schedule could be established similar to that developed by Wehrmann and Scharpf [26]. On soils rich in vermiculite the non-exchangeable NH_4^+ should be considered also for fertilizer N recommendation in lowland rice.

References

1. Broadbent FE and Nakashima T (1970) Nitrogen immobilization in flooded soils. Soil Sci Soc Amer Proc 34:218–221
2. Broadbent FE and Tusneem FE (1971) Losses of nitrogen from flooded soils in tracer experiments. Soil Sci Soc Amer Proc 35:922–926
3. Bremner JM (1965) Inorganic forms of nitrogen. In C.A. Black ed Methods of soil analysis. Amer Soc Agron, Madison, Wisconsin: 1179–1237
4. Dahnke WC and Vasey EH (1973) Testing soils for nitrogen in Walsh LM and JD Beaton Soil testing and plant analysis. Soil Sci Soc Amer, Madison, Wisconsin:97–115
5. Faust H (1969) Optical spectroscopy technique for N-15 assay. IAEA/FAO. Res. coord. meeting on recent development in the use of N-15 in soil-plant studies, Sofia, Bulgaria
6. Fox RH and Pickielek WP (1978) A rapid method for estimating the nitrogen-supplying capacity for a soil. Soil Sci Soc Amer J 42:751–753
7. Harwood JE and Kuhn AL (1970) A colorimetric method for ammonia in natural waters. Water Res 4:805–811
8. Keeney DR and Bremner JM (1966a) A chemical index of soil nitrogen availability. Nature, 211:892–893
9. Keeney DR and Bremner JM (1966) Determination and isotope ratio analysis of different forms of nitrogen in soils. 4. Exchangeable ammonium, nitrate, and nitrite by direct distillation methods. Soil Sci Soc Amer Proc 30:583–594
10. Keerthisinghe G, Mengel K and De Datta SK (1984) The release of non-exchangeable ammonium (^{15}N labelled) in wetland rice soils. Soil Sci Soc Amer J 48:291–294
11. Kerbs LD, Jones JD, Thiessen WL and Parks FP (1973) Correlation of soil test nitrogen with potato yields. Comm. in Soil Science and Plant Analysis 4:269–278
12. Kowalenko CG and Ross GJ (1980) Studies on the dynamics of 'recently' clay fixed NH_4^+ using ^{15}N. Can J Soil Sci 60:61–70
13. Martin AE, Gilkes RJ and Skjeemstad JO (1970) Fixed ammonium in soils developed on some Queensland phyllites and its relation to weathering. Aust J Soil Res 8:71–80
14. Mengel K and Scherer HW (1981) Release of non-exchangeable (fixed) soil ammonium under field conditions during the growing season. Soil Sci 131:226–232
15. Mohammed IH (1979) Fixed ammonium in Libyan soils and its availability to barley seedlings. Plant and Soil 53:1–9
16. Opuwaribo E and Odu CTI (1974) Fixed ammonium in Nigerian soils. I. Selection of a method and amounts of native fixed ammonium. J Soil Sci 25:256–264
17. Savant NK and De Datta SK (1982) Nitrogen transformations in wetland rice soils. Adv Agron 35:241–302
18. Schön HG (1982) Die Bedeutung des austauschbaren Ammoniums in überfluteten Reisböden für die Ertragsbildung von Reis und für die Basis einer Stickstoffdünger-empfehlung. Ph D thesis, Fac of Nutrition, Justus Liebig-University, Giessen
19. Scott AD and Smith SJ (1966) Susceptibility of interlayer potassium in micas to exchange with sodium. Clays and Clay Min Proc 14th Nat Conf 69–81
20. Silva JA and Bremner JM (1966) Determination and isotope-ratio analysis of

different forms of nitrogen in soils. 5. Fixed ammonium. Soil Sci Soc Amer Proc 30:587–594

21. Sims JR and Jackson GD (1971) Rapid analysis of soil nitrate with chromotropic acid. Soil Sci Soc Amer Proc 35:603–606
22. Sippola J, Ervio R and Eleveld R (1973) The effect of simultaneous addition of ammonium and potassium on their fixation in some Finnish soils. Ann Agriculturae Fenniae 12:185–189
23. Smith JA (1966) An evaluation of nitrogen soil test methods for Ontario soils. Can J Soil Sci 46:185–194
24. Soper RJ and Huang RM (1963) The effect of nitrate nitrogen in the soil profile on the response of barley to fertilizers nitrogen. Can J Soil Sci 43:350–358
25. Walsh LM and Murdock JT (1963) Recovery of fixed ammonium by corn in greenhouse studies. Soil Sci Soc Amer Proc 27:200–204
26. Wehrmann J and Scharpf HC (1979) Der Mineralstoffgehalt des Bodens als Maßstab für den Stickstoffdüngerbedarf (N_{min}-Methode). Plant and Soil 52:109–126

7. The efficacy and loss of fertilizer N in lowland rice

PAUL LG VLEK and BERNARD H BYRNES[1]

Agro-Economic Division, International Fertilizer Development Center (IFDC),
P.O. Box 2040, Muscle Shoals, Alabama 35662, USA

Key words: fertilizer efficiency, ^{15}N, nitrogen losses, flooded soils

Abstract. Nitrogen fertilization is a key input in increasing rice production in East, South, and Southeast Asia. The introduction of high-yielding varieties has greatly increased the prospect of increasing yields, but this goal will not be reached without great increases in the use and efficiency of N on rice. Nitrogen enters a unique environment in flooded soils, in which losses of fertilizer N and mechanisms of losses vary greatly from those in upland situations. Whereas upland crops frequently use 40–60% of the applied N, flooded rice crops typically use only 20–40%. There is a great potential for increasing the efficiency of N uptake on this very responsive crop to help alleviate food deficits in the developing world.

This article reviews current use of N fertilizers (particularly urea) on rice, the problems associated with rice fertilization, and recent research results that aid understanding of problems associated with N fertilization of rice and possible avenues to increase the efficiency of N use by rice.

Approximately 40% of the world's population depend on rice (*Oryza sativa* L.) as their major caloric source, and in many of the less-developed countries of Asia, 80 to 90% of the population rely on rice as their staple food [11]. A native of monsoonal Asia, rice has been spread into many diverse environments, both in Asia and elsewhere. On a world basis Asia accounts for about 90% of the total area cultivated in rice [11]. Rice is unique in that it is the only major food crop that is semiaquatic, that grows best in a flooded soil. This preferred habitat is a major source of difficulty in the maintenance of nitrogen (N) added as chemical fertilizers. This paper attempts to review (1) the importance of N fertilizers in increasing rice production and (2) recent research findings related to N fertilization of rice.

The rice environment

The environment in which rice is grown varies enormously. Rice is found as far north as Hokkaido, Japan, and as high as 3000 m in Nepal. It is raised successfully in desert areas of the Middle East and Peru and in the tidal

---

[1] Director, Agro-Economic Division, and Research Associate, respectively.

131

swamps of major river floodplains such as the Niger or Brahmaputra. However, except for the latter environment, rice would not be found in these places without human intervention. The climatic conditions in which rice naturally occurs are very warm and humid, with strong monsoonal influence and low indirect solar radiation [22].

Although by origin and preference a lowland crop [34], rice can be grown as an upland crop through varietal adaptation. Upland rice in freely drained fields is restricted to areas with adequate rain and probably constitutes less than 10% of all rice lands. Irrigated rice dominates in China and subtropical areas, whereas upland rice is most common in Africa and Latin America. Rainfed lowland rice is predominant in South and Southeast Asia, accounting for probably 75% of the total planted [11]. The majority of the studies on the fate of fertilizer N under rice have concentrated on transplanted, irrigated rice, in both the tropics and temperate zones.

Fertilizer applied to lowland rice enters a unique flooded field soil-plant-water-atmosphere system which was first described by Pearsall [43]. The soil is overlaid by floodwater, which restricts movement of oxygen into the soil. The surface layer of soil is somewhat oxidized to a thickness which varies from 0 to 3 cm, depending on oxygen concentrations in the floodwater and the rates of oxygen consumption [21] or production [19]. To maintain respiration under water, rice has highly developed aerenchyma which allow the plant to develop an oxidized rhizosphere, the importance of which to the N economy is still in question [2, 18]. Once entered into this system, fertilizer N (urea or ammonium salts) is subject to a complex set of processes (Figure 1) discussed elsewhere [8] and in this issue. The net result of these transformations and processes is reflected in the response and fertilizer use by the crop.

Fertilizer N applications for rice are principally designed to complement the N available from soil organic matter, including plant residues, manure additions, and biologically fixed N. The amounts of N derived from sources other than fertilizer vary with environment, season, time, and crop management [9]. Although vast amounts of N are potentially available in organic wastes in rice-growing countries, the extent to which these are used varies greatly [55]. In China, where use of organic manures has been exemplary, the recently installed capacity of nearly 7 million mt of N per annum [51] will probably result in greater reliance on fertilizer N for rice production.

Rice production in the tropics has been increased through the combined adoption of high-yielding varieties (HYV), fertilizers, and chemical crop protection. Progress in this area was rapidly promoted by the development of short-statured, stiff-strawed, photoperiod-insensitive HYV in the mid-sixties The cost of adopting the HYV was attractive in regions where rice could be produced under reliable rainfall or irrigation. However, the areas with more risky environments for rice production, e.g., deepwater and upland rainfed rice regions, have been largely bypassed by the 'Green Revolution'.

Recent increases in Asian rice production have been impressive. From

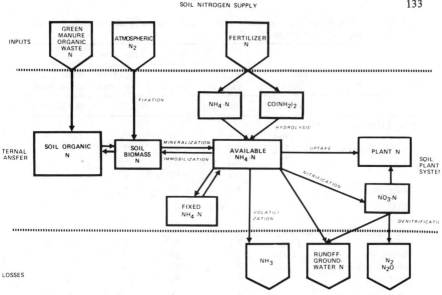

Figure 1. Nitrogen inputs, losses, and internal transfers in the rice-submerged soil system

1965–75, rice production in Asia increased by approximately 2.8%/year. Approximately 60% of the gained production is attributed to yield increase through more intensive cultivation. Of this increase, 62% is attributed to the increased use of fertilizer. The remaining 38% has been due to the use of HYV and other factors, including chemicals for weed and insect control and irrigation development. High yields of rice are directly related to use of modern varieties, fertilizers, and good water control. Lack of these modern inputs restricts yields at 1–1.5 mt/ha, whereas the combined use of these inputs may increase yields to between 5 and 6 mt/ha. Nitrogen accounts for the largest fraction of fertilizer applied to rice (67%). During 1980/81, rice accounted for approximately 4.3 million mt of N consumption in Asia. Assuming an agronomic efficiency of 10 kg rice/kg of N, this would account for 43 million mt of rice or 10% of the world rice production.

Some relevant fertilizer statistics for Asia are summarized in Table 1. The wide discrepancy in the use of fertilizer N among countries can be largely explained by the two factors of (1) production risk and (2) the cost of rice relative to the cost of N fertilizers. Low average use of fertilizer in a country generally reflects the limited availability of land where rice production is relatively free of risk. For instance, a detailed farm survey conducted by Sidhu et al. [48] in Bangladesh indicated that in 1980/81 fertilizer use was 69 kg/ha, distributed over one or more seasons. The three-season average use of fertilizer for 1979/80 for traditional rice varieties was 41 kg/ha, whereas high-yielding varieties received an average of 193 kg/ha. Although only 30% of the total rice area was used for HYV, the HYV consumed 66% of the fertilizer applied to rice. The three-season average fertilizer use per cropped

Table 1. Estimates of rice area harvested, type of water regime, percent modern varieties, yield, and average use of nitrogen for selected areas

Area	Harvested Rice area	Rice yield	Harvested rice area		Average N use based on arable land
			Irrigated	Modern varieties	
	$(10^3$ ha$)$	(mt/ha)	(%)		(kg N/ha)
Group I (high rice yields)					
Japan	2764	5.5	94	100	149
Korea (Republic of)	1218	5.9	85	90	209
Taiwan	787	5.2	83	95	149
Egypt	442	5.2	100	85	152
Average	–	5.5	91	93	165
TOTAL	5211	(28778)	–	–	–
Group II (modern varieties rice yields)					
China	35390	3.2	76	80	32
Indonesia	8369	2.6	58	40	26
Malaysia (West)	585	2.8	48	38	115
Iran	461	3.5	90	NA	12
Average	–	3.0	68	72	46
TOTAL	44805	(138259)	–	–	–
Group III (low yields)					
India	39688	1.8	43	25	13
Pakistan	1710	2.3	80	43	38
Bangladesh	10329	1.9	5	14	16
Philippines	3579	1.8	45	56	28
Vietnam	5310	2.2	16	17	36
Thailand	8383	1.8	37	5	12
Burma	5069	1.8	16	6	4
Sri Lanka	597	2.0	66	60	42
Afghanistan	210	2.1	6	NA	3
Nepal	1256	1.9	10	19	4
Laos	680	1.2	20	NA	NA
Kampuchea	555	1.3	3	NA	NA
Average	–	1.8	29	27	20
TOTAL	77366	(142892)	–	–	–
Total for 20 countries					
Average	–	2.4	50	42	23
TOTAL	127382	(309929)	–	–	–
WORLD TOTAL	142085	2.4	NA	NA	31

Note: Values in parentheses are total rice production (mt). Adapted from Stangel [51]

area of HYV was approximately 43% higher under irrigated conditions than under rainfed conditions. As a result, the average fertilizer use per area cropped and fertilized was highest (219 kg/ha) during the Boro (dry) season when irrigation and HYV are most common [48]. Thus, fertilizer use generally goes hand in hand with the use of other inputs.

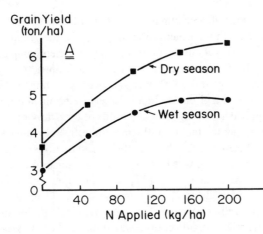

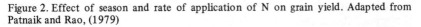

Figure 2. Effect of season and rate of application of N on grain yield. Adapted from Patnaik and Rao, (1979)

Figure 2 demonstrates the effect of dry- versus wet-season environmental conditions on grain yield response to N of an HYV (Jaya) in India [40]. The decreased yield response in the wet season is a result of the reduced solar radiation in combination with high relative humidity and increased disease and insect incidence. The effect of complementary inputs was demonstrated by David and Barker [10] for farmers' field conditions in the Philippines. They showed the benefits of fertilizer N to be largely restricted to improved varieties (HYV). However, without applied N the benefits of HYV are nil [52].

Fertilizer efficiency

Efficient use of N applied to rice has been a cause for concern for some time. In field experiments, flooded rice generally recovers only 20–40% of applied N [33], whereas upland crops normally recover about 40–60%. There is little reason to believe that farmers' fertilization practices are performing any better. However, data are lacking to make firm conclusions. The International Fertilizer Development Center (IFDC), in collaboration with the International Rice Research Institute, established a joint project in 1976 to assess the efficiency of use of applied N as a function of fertilizer source or management. Various national programs in Asia joined this effort. Results from these studies and information available in the literature are reviewed here.

The concept of fertilizer efficiency is often poorly understood; therefore, the term 'efficiency' should be carefully defined before it is used. One of the important parameters is the agronomic efficiency (AE), i.e., the response in yield per unit of input. However, the dependence of AE on the fertilizer application rate precludes comparisons of fertilizers applied at different rates. To circumvent this problem, one may express the AE at a single application rate or, better yet, use regression analysis when comparing fertilizer response [53]. The performance of applied N is known to depend on the source and management of the fertilizer in a given environment [58].

The performance of a given product is largely determined by the efficiency of use of the applied N during critical periods of crop development [32, 59], and many environmental factors influence the efficiency of use of fertilizer N [14]. The efficiency may be affected by numerous crop- and water-management factors such as weeding, transplanting versus direct seeding [44], plant density and plant geometry [37], and flooding regime [45, 50]. The uniformity and method of incorporation of fertilizer are particularly important, and it is disconcerting that many reports on experiments do not even mention whether incorporation was done or how it was done. Thus evaluating the performance of fertilizers and fertilizer management practices is often a difficult task.

Experiment station grain yields commonly exceed 4 mt/ha when conventional fertilizer practices are used, and in many instances alternative practices or N sources will produce 10–20% higher yields than conventional practices. With coefficients of variation of the experiments of 10% or more, such differences are often not statistically significant. Yet, such yield improvements, if they are real, are of interest to the farmer. Thus, researchers concerned with fertilizer management should consider experimental designs that allow more precise conclusions to be drawn. Increasing the number of replicates and conducting multirate (e.g., four nonzero rates) experiments [53] are two approaches to decreasing statistical variability. Verification of research results under farm-level conditions is an essential part of the evaluation process, and care in site selection and replication is needed, along with supporting soil and climatic data, to better evaluate response to fertilizers.

The problems associated with high variability in fertilizer yield response have led scientists to evaluate the recovery of fertilizer N by the crop. Traditionally, fertilizer recovery is calculated as the differential N uptake between fertilized and nonfertilized crop per quantity of applied N. For instance, in an experiment comparing ammonium nitrate (AN) and urea for rice, Sharma and Ghosh [47] reported yields of 3.36, 5.71, and 5.22 mt/ha for the check, urea, and AN treatments, respectively. The yield advantage of urea over AN was 9% which was barely significant at the 5% level. More convincingly, the N uptake was 16% higher from urea than from ammonium nitrate while apparent N recovery increased by 38%. Although there are no means of ascertaining whether the increased uptake of N is indeed derived from the fertilizer, the increased uptake is definitely caused by the applied N. Additional uptake of soil N may occur because of stimulated mineralization of soil N (priming effect) or simple mineralization-immobilization turnover

[26], combined with a healthier plant's increased demand for N [49] and better developed root system to scavenge N from the soil.

The stable isotope ^{15}N has been used in recent studies in an attempt to gain additional insight into the use of fertilizer N. Analysis of the ^{15}N content in the plant at selected times during the growing season allows calculation of actual fertilizer recovery at various stages of development. However, in many cases the true effect of fertilizer may be over- or underestimated because of mineralization-immobilization turnover. The disparity between recovery calculated with and without the use of ^{15}N for a range of rice experiments is shown in Figure 3, which is taken from Vlek and Fillery [58]. Generally, on relatively N-fertile soils, ^{15}N recovery tends to be higher than apparent recovery, because of the cycling of the labeled isotope into the large soil N pool. For soils low in native N, ^{15}N recovery will be in closer agreement with apparent recovery or may be lower in case immobilization of fertilizer nitrogen occurs to a significant extent.

Craswell and Vlek [8] summarized published literature on experiments in which ^{15}N was used to calculate fertilizer recovery (Table 2). Recoveries of ^{15}N ranged from as low as 7% to as high as 68%. However, as is clear from Figure 3, these recoveries reveal only part of the effect of fertilizer uptake on nitrogen use.

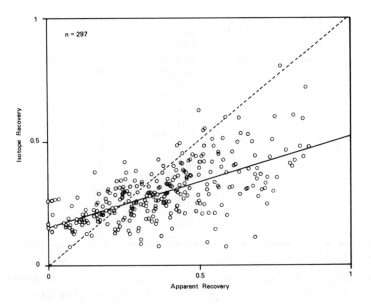

Figure 3. Relation between apparent recovery and ^{15}N recovery by rice. Composite plot of data points derived from a series of experiments coordinated by IAEA, 1976

138

Table 2. Recovery by rice of ^{15}N-labeled fertilizers

Type of experiment	Soil or location	Fertilizer Material[a]	Placement and timing	[15]N recovery in plant (%)	References
Pot	–	AS	Basal	38	[39]
			2/3 planting	40	
			1/3 boot		
Field	Sri Lanka	AS	Surface	11	[36]
			5 cm deep	20	
Field	2 Philippines soils	AS, U	Surface	28	[13]
			Deep	68	
			Split	34	
Pot			Surface	50	[13]
			Deep	66	
			Split	47	
Field	Maahas		Basal	35	[13]
		AS	Best split	45	
	Maligaya		Basal	18	
			Best split	47	
Field	Thailand	AS	Basal	17–42	[29]
			Split	40	
			Deep	18	
			Surface	8	
			+ Nitrification inhibitor	12	
			Flowering	77	
Field	USA	AS		17–23	[41]
Field	India	AS, U	Basal	11	[54]
			Panicle initiation	27	
Field	India	AS, U	Surface	18	[28]
			Incorporation	29	
			Deep	38	
			Heading	37	
Field	USA	AS	Deep	48	[42]
			Early	38–51	
			Midseason	33	
			Split	35–61	
Field	Japan		Surface	23	[35]
			Incorporation	48	
			Basal	46	
			Split	47	
			Deep	63	
			Basal	7	
			Heading	49	

[a] AS = ammonium sulfate, U = urea

In many of the IAEA field experiments, rice plants were harvested and analyzed for ^{15}N so that the percentage of N in the plant derived from the fertilizer (Ndff) could be calculated [23, 24]. In many of the rice experiments, deep placement increased the Ndff values above those for surface application; ammonium sulfate and urea behaved similarly and were superior to sodium nitrate. Discussing the value of such data, Rennie and Fried [46] suggested that the Ndff parameter provides a sensitive criterion by which to assess specific fertilizer practices but should be used only for comparisons within a site. However, Jansson and Persson [26] argue that ^{15}N recoveries by the crop are largely misleading and should only be interpreted in the context of the total ^{15}N balance in the soil-plant system.

Loss of fertilizer N

Labeled N fertilizers have been used successfully in research on N losses in lowland rice. As a result it is now widely appreciated that the major cause of the inefficient use of N by rice is the extensive loss of N which accompanies the poor management of N fertilizer. Craswell and Vlek [8] have summarized the earlier literature on ^{15}N loss which was obtained largely from greenhouse pot experiments. In this review we will focus on recent reports of ^{15}N loss from field studies. Data were available from experimental sites at two locations in the Philippines, northern India (Punjab), Thailand (Bangkok Plain), Korea (Suweon), eastern China (Nanjing), and southern China (Fujian), covering both tropical and subtropical environments. A summary of fertilizer N budgets recorded at these sites is provided in Table 3 for the two most commonly used N sources: urea and ammonium sulfate. With few exceptions fertilizer losses were estimated to be in excess of 35%. The cause of these losses is discussed elsewhere [58].

The magnitude of fertilizer N loss is related to soil and climatic factors and fertilizer management practices. Generally, the highest losses of N occur from broadcast of urea into the floodwater at an early plant growth stage. For example, Fillery and De Datta [15] and Fillery et al. [16] showed losses of 40—60% of the ^{15}N-labeled urea or ammonium sulfate broadcast to floodwater 2—4 weeks after transplanting — common farmers' practice in the Philippines. The broadcast and incorporation of urea before transplanting of rice seedlings resulted in lower ^{15}N losses in some studies but losses comparable to those obtained with farmers' management in others (Table 3). It is hypothesized that the disparity in results obtained with broadcast and incorporation of fertilizer N reflects, in large measure, the array of incorporation techniques used. Soil conditions and the depth of floodwater at the time of fertilizer application may also affect the efficiency of incorporation. In most fertilizer evaluation experiments for rice, alternative sources or techniques have been compared with broadcast and incorporation of N. Inconsistent incorporation may at least in part account for the variable increases in grain yield often noted for the alternative N sources.

Fertilizer N remaining in the soil was generally present in the top 15 cm. Even in the most permeable soil (India), 17—20% of the applied ^{15}N was recovered in the 0—15 cm layer and less than 4% from the 15—30 cm depth.

Table 3. Summary of ^{15}N balance sheets recently reported for conventional N management in transplanted rice

Location	Season	References	N source	Rate (kg/ha)	Management[b]	N budget[a]		
						Plant	Soil	Loss
						(%)		
Philippines, Los Baños	Dry	[25]	Urea[c]	60	AT	9.0	31.0	60.0
		[17]	Urea	87	AT	35.4	42.3	22.0
		[25]	Urea[c]	80	BI	1.0	76.0	23.0
		[4]	Urea	87	BI	33.0	56.0	11.0
		[7]	Urea	58	BI	23.0	15.9	61.0
		[7]	Ammonium sulfate	58	BI	22.8	11.6	65.0
	Wet	[4]	Urea	58	BI	32.0	33.0	35.0
		[7]	Urea	38	BI	36.6	22.1	41.0
		[7]	Ammonium sulfate	38	BI	27.3	17.8	55.0
—, Munoz	Dry	[25]	Urea[c]	80	AT	2.0	51.0	47.0
		[15]	Urea[c]	58	AT	10.9	46.9	42.0
		[17]	Urea[c]	87	AT	34.6	39.0	26.0
		[25]	Urea[c]	60	BI	1.0	81.0	18.0
		[15]	Ammonium sulfate[c]	58	AT	11.3	42.6	44.0
China, Nanjing	—	[5]	Urea	59	BI	29.0	31.5	39.5
		[5]	Urea	59	SB	22.3	30.4	47.3
		[5]	Urea	59	BS	37.2	22.7	40.1
		[5]	Urea	59	SB	27.5	18.6	53.9
		[5]	Urea	59	BS	44.9	19.8	35.3
		[5]	Urea	59	SB	39.8	16.4	43.8
—, Fujian	—	(Liu et al.)[d]	Urea	40	BI	23.2	18.8	58.0
		(Liu et al.)[d]	Urea	80	BI	20.2	19.8	60.0
		(Liu et al.)[d]	Ammonium sulfate	40	BI	21.4	24.8	53.8
		(Liu et al.)[d]	Ammonium sulfate	80	BI	24.5	18.3	57.2

Table 3. (*Continued*)

Location	Season	References	N source	Rate (kg/ha)	Management [b]	N budget [a] Plant (%)	Soil (%)	Loss (%)
Korea, Suweon	—	(Huh et al.)[d]	Urea	40	BI	32.7	53.5	13.8
		(Huh et al.)[d]		80	BI	39.4	35.4	25.2
		(Huh et al.)[d]	Ammonium sulfate	40	BI	36.5	54.0	9.5
		(Huh et al.)[d]		80	BI	54.2	35.2	10.6
Thailand, Bangkhen	—	(Snitwongse et al.)[d]	Urea	20	BI	17.4	52.7	29.9
		(Snitwongse et al.)[d]		40	BI	48.0	30.3	21.7
		(Snitwongse et al.)[d]	Ammonium sulfate	20	BI	24.1	63.6	12.3
		(Snitwongse et al.)[d]		40	BI	28.9	47.2	23.9
		(Snitwongse et al.)[d]	Ammonium chloride	20	BI	20.1	60.8	19.1
		(Snitwongse et al.)[d]		40	BI	27.2	51.3	21.5
India, Ludhiana		[27]	Urea	40	BI	21.0	29.0	50.0
		[27]		80	BI	26.0	25.0	49.0
		[27]	Ammonium sulfate	40	BI	23.0	31.0	46.0
		[27]		80	BI	31.0	21.0	48.0

[a] At harvest unless otherwise specified
[b] AT = after transplanting; BI = broadcast and incorporated; SB = split broadcast; BS = basal surface applied
[c] Determined during course of experiment
[d] Unpublished results from IFDC-sponsored research with Dr. Liu Chung Chu, Fujian Academy of Sciences China; Dr. Beom Lyang Huh, Kangweon Provincial Office of Rural Development, Korea; Dr. P. Snitwongse, Department of Agriculture, Bangkhen, Thailand

Negligible quantities of ^{15}N were found below 30 cm at the end of the growing season [27]. Most of the fertilizer [15]N ($>98\%$) remaining in the soil was incorporated into the organic pool by final harvest [7].

Plant recoveries of fertilizer N by the crop at harvest vary from 17.4 to 54.2%. Low plant recoveries generally go hand in hand with high loss of applied N. However, where this was not the case (Thailand and Korea), low efficiency was due to high rates of immobilization. In these countries the crops may actually have benefited from the fertilizer N more than is apparent, due to biological interchange [26].

Deep placement of fertilizer has proved to be an effective means of reducing ammonia concentrations in the floodwater [6] and thus reducing the potential for ammonia volatilization [57]. Simultaneously, deep placement of urea decreases the conversion of ammonium to nitrate and thereby reduces denitrification losses. Regardless of whether ammonia volatilization or denitrification is responsible for the poor efficiency of urea in rice, the efficacy of deep placement has been confirmed, both in ^{15}N experiments [23, 1, 3, 38, 8, 30, 4, 7] and in simple source comparison trials [12, 20, 31]. A summary of the performance of urea supergranules (USG) in trials conducted by various organizations in association with IFDC is presented by Vlek and Fillery [58].

Deep placement of urea or ammonium bicarbonate has drastically reduced N loss in all field studies conducted on soils with low percolation rates (Figure 4); however, measurable N loss still occurred in several studies in the Philippines and China. Deep placement of urea in soils subject to high percolation rates (Punjab, India) sharply increased N loss as compared to

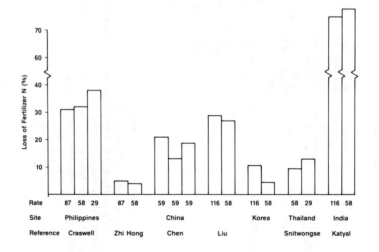

Figure 4. [15]N losses from deep-placed urea supergranules

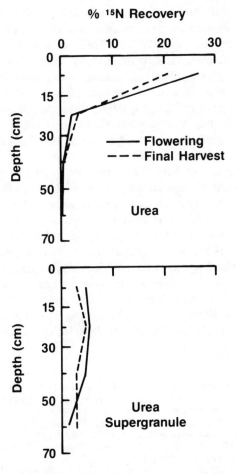

Figure 5. Distribution of ^{15}N applied as broadcast incorporated urea and urea super-granule to rice in Punjab, India

broadcast and incorporation, primarily because leaching of N was increased [27]. In the latter study, losses from USG averaged 60–78% and were significantly different from those of the broadcast and incorporated urea. The distribution of ^{15}N urea in this loamy sand at harvest time for both sources is illustrated in Figure 5. High leaching rates (> 10 cm/day) combined with high localized concentrations of N are the likely cause for the loss of ^{15}N from the USG-fertilized soil [56].

Summary and conclusions

Rice is an increasingly important crop, presently feeding over 40% of the world population. In the early 1960s large areas in Southeast Asia, the rice bowl of the world, were experiencing rice deficits. With the introduction of high-yielding rice varieties and the commitment to fertilizer use, the region has gradually become self-sufficient. Total consumption of fertilizer N on rice to date has reached 3 million mt annually, over 90% of which is urea.

The efficiency of fertilizer use is low, ranging from 20 to 40% recovery by the crop, depending on N source, management, and agroecological conditions. The poor efficiency is largely due to loss of fertilizer N. Nitrogen recoveries can be improved substantially by improved N sources or management, and this improvement would obviously result in enormous savings to developing economies and farmers alike.

At current prices of US$200/mt of urea, an improvement in efficiency of use of applied N through elimination of N loss will save in excess of US$16 million annually for each percentage point. Assuming that fertilizer efficiency can be doubled as a result of fertilizer or fertilizer management improvement, annual savings may reach as high as US$300 million per year.

The use of ^{15}N balance techniques has clearly identified N loss as a major problem in lowland rice management. Such studies also provide considerable information on the time frame and thus the mechanisms causing N loss. Since this information is invaluable to research programs concerned with the improvement of N use by rice, considerable attention has to be given to the role of different loss mechanisms in different agroecological settings.

References

1. Aleksic Z, Broeshart H and Middleboe V (1968) Shallow depth placement of $(NH_4)_2SO_4$ in submerged rice soils as related to gaseous losses of fertilizer nitrogen and fertilizer efficiency. Plant Soil 28:338–342
2. Bouldin DR (1966) Speculation on the oxidation-reduction status of the rhizosphere of rice roots in submerged soils. FAO/IAEA Tech Rep 65:128–139
3. Broadbent FE and Mikkelsen DS (1968) Influence of placement on uptake of tagged nitrogen by rice. Agron J 60:674–677
4. Cao ZH, De Datta SK and Fillery IRP (1984) Nitrogen-15 balance and residual effects of urea-N in wetland rice fields as affected by deep-placement techniques. Soil Sci Soc Am J 48:203–208
5. Chen RY and Zhu ZL (1982) Characteristics of the fate and efficiency of nitrogen in supergranules of urea. Fert Res 3:63–72
6. Craswell ET, De Datta SK, Obcemea WN and Hartantyo M (1981) Time and mode of nitrogen fertilizer application to tropical wetland rice. Fert Res 2:247–259
7. Craswell ET, De Datta SK, Weeraratne CS and Vlek PLG (1985) Fate and efficiency of nitrogen fertilizers applied to wetland rice. I. The Philippines. Fert Res 6(1):49–63
8. Craswell ET and Vlek PLG (1979) Fate of fertilizer nitrogen applied to weland rice. In: Nitrogen and Rice, pp. 175–192. International Rice Research Institute, Los Baños, Philippines
9. Craswell ET and Vlek PLG (1982) Nitrogen management for submerged rice soils. In: Proceedings 12th International Congress of Soil Science, New Delhi 3:158–181
10. David C C and Barker R (1978) Modern rice varieties and fertilizer consumption. In: Economic consequences of the new rice technology. International Rice Research Institute, Los Baños, Philippines
11. De Datta SK (1981) Principles and practices of rice production. John Wiley and Sons, Inc., New York. 618 p

12. De Datta SK and Gomez KA (1981) Interpretive analysis of the international trials on nitrogen fertilizer efficiency in wetland rice. Fertilizer International. Pages 1–5. The British Sulphur Corporation, Ltd., London, UK
13. De Datta SK, Magnaye CP and Moomaw JC (1968) Efficiency of fertilizer nitrogen (15N-labeled) for flooded rice. Int Congr Soil Sci, Trans. 9th (Adelaide, Aust.) 4:67–76
14. De Datta SK and Zarate PM (1970) Environmental conditions affecting the growth characteristics, nitrogen response, and grain yield of tropical rice. Biometeorology 4(1):71–89
15. Fillery IRP and De Datta SK (1985) Effect of N source and a urease inhibitor on NH_3 loss from flooded rice. (1) Ammonia fluxes and ^{15}N loss. Soil Sci Soc Am J (submitted)
16. Fillery IRP, De Datta SK and Craswell ET (1985) Effect of phenyl phosphoro-diamidate on the fate of urea applied to wetland rice fields. Fert Res (in press)
17. Fillery IRP, Simpson JR and De Datta SK (1985) Contribution of ammonia volatilization to total N loss after applications of urea to wetland rice fields. Fert Res (in press)
18. Fillery IRP and Vlek PLG (1982) The significance of denitrification of applied nitrogen in fallow and cropped rice soils under different flooding regimes. Plant Soil 65:153–169
19. Fillery IRP and Vlek PLG (1985) Reappraisal of the significance of ammonia volatilization as a N loss mechanism in flooded rice fields. Fert Res (this issue)
20. Flinn JC, Mamaril CP, Velasco LE and Kaiser K (1984) Efficiency of modified urea fertilizers for tropical irrigated rice. Fert Res 5:157–174
21. Howeler RH and Bouldin DR (1971) The diffusion and consumption of oxygen in submerged soils. Soil Sci Soc Am Proc 35:202–208
22. Huke R (1976) Geography and climate of rice. pp. 31–50, in Climate and rice. International Rice Research Institute, Los Baños, Philippines
23. International Atomic Energy Agency (IAEA) (1970) Rice fertilization. A six-year isotope study on nitrogen and phosphorus fertilizer utilization. IAEA Tech Rep Ser 108:1–184
24. IAEA (1976) N-15 labelled fertilizer studies on flooded rice, mimeographed (52 p. plus Appendixes)
25. International Fertilizer Development Center (IFDC) (1983) Annual Report, IFDC, Muscle Shoals, Alabama 35662, USA
26. Jansson SL and Persson J (1982) Mineralization and immobilization of soil nitrogen. In Stevenson FJ (ed) Nitrogen in agricultural soils, p. 229–252. American Society of Agronomy, Madison, WI, USA
27. Katyal JC, Bijay Singh, Vlek PLG and Craswell ET (1985) Fate and efficiency of nitrogen fertilizers applied to wetland rice. II. Punjab, India. Fert Res 6: 279–290
28. Khind, CS and NP Datta (1975) Effect of method and timing of nitrogen application on yield and fertilizer nitrogen utilization by lowland rice. J Indian Soc Soil Sci 23:442–446
29. Koyama T, Chamnek C and Niamrischand N (1973) Nitrogen application technology for tropical rice as determined by field experiments using ^{15}N tracer technique. Trop Agric Res Cent Tokyo, Japan Tech Bull TARC 3:1–79
30. Li CK and Chen RY (1980) Ammonium bicarbonate used as a nitrogen fertilizer in China. Fert Res 1:125–136
31. Martinez A, Diamond RB and Dhua SP (1983) Agronomic and economic evaluation of urea placement and sulfur-coated urea for irrigated paddy in farmers' fields in Eastern India. Paper Series IFDC-P-4, International Fertilizer Development Center, Muscle Shoals, AL, USA
32. Matsushima S (1975) Crop science in rice. Fuji Publ Co, Ltd, Tokyo, Japan
33. Mitsui S (1954) Inorganic nutrition, fertilization, and soil amelioration for lowland rice. Yokendo Ltd, Tokyo, Japan
34. Moormann FR and van Breemen N (1978) Rice: soil, water, land. International Rice Research Institute, Los Baños, Philippines

35. Murayama N (1979) The importance of nitrogen for rice production. Nitrogen and Rice, pp. 5–23. IRRI, Los Baños, Philippines
36. Nagarajah S and Al-Abbas AH (1967) Nitrogen and phosphorus fertilizer placement studies on rice using N^{15} and P^{32}. Trop Agric 121:89–103
37. Ngure NV and De Datta SK (1979) Increasing efficiency of fertilizer nitrogen in wetland rice by manipulation of plant density and plant geometry. Field Crops Research 2:19–34
38. Obcemea WN, De Datta SK and Broadbent FE (1984) Movement and distribution of fertilizer nitrogen as affected by depth of placement in wetland rice. Fert Res 5:125–148
39. Patnaik S (1965) N^{15} tracer studies on the utilization of fertilizer nitrogen by rice in relation to time of application. Proc. Indian Acad Sci Sect B 61:31–39
40. Patnaik S and Rao MV (1979) Sources of nitrogen for rice production. In Nitrogen and rice, pp. 25–43. International Rice Research Institute, Los Baños, Philippines
41. Patrick WH Jr, Delaune RD and Peterson FJ (1974) Nitrogen utilization by rice using ^{15}N-depleted ammonium sulfate
42. Patrick WH Jr and Reddy KR (1976) Fate of fertilizer nitrogen in a flooded rice soil. Sci Soc Am J 40:678–681. Agron J 66:819–820
43. Pearsall WH (1938) The soil complex in relation to plant communities. I. Oxidation-reduction potentials in soils. J Ecol 26:180–193
44. Rao MV, Prasad M and Rao SS (1975) Management of low rates of nitrogen for rice. Fert News 20:31–32
45. Rajale GB and Prasad R (1975) Nitrogen and water management for irrigated rice. Riso 24:117–125
46. Rennie DA and Fried M (1971) An interpretative analysis of the significance in soil fertility and fertilizer evaluation of ^{15}N labelled fertilizer experiments conducted under field conditions. pp. 639–656. In International Symposium Soil Fertility Evaluation Proceedings Vol. 1. Indian Society of Soil Science, New Delhi, India
47. Sharma RC and Ghosh AB (1977) Fertilizer efficiency of ammonium nitrate for rice. Fert News 22:27–29
48. Sidhu SS, Baanante CA and Ahsan E (1982) Agricultural production fertilizer use and equity considerations. Results and analysis of farm survey data, 1979/80, Bangladesh. International Fertilizer Development Center, Muscle Shaols, AL, USA
49. Sinclair TR and DeWit CT (1975) Photosynthate and nitrogen requirements for seed production by various crops. Science 189:565–567
50. Singh A and Pal RA (1973) A note on the response of transplanted rice to nitrogen and water management practices. Ind J Agron 18:376–377
51. Stangel PJ (1979) Nitrogen requirement and adequacy of supply or rice production. In Nitrogen and rice, pp. 45–69. International Rice Research Institute, Los Baños, Philippines
52. Te A and Flinn JC (1984) New rice technology and fertilizer demand. FADINAP/GPA/IRRI Regional Workshop on Fertilizer Demand Projections, Manila, March 5–9, 1984
53. Tejeda HR, Hong CW and Vlek PLG (1980) Comparison of modified urea fertilizers and estimation of their availability coefficient using quadratic models. Soil Sci Soc Am J 44:1256–1262
54. Upadhya GS, Datta NP and Deb DL (1974) Note on the effect of selected drainage practices on yield of rice and the efficiency of nitrogen use. Indian J. Agric Sci 43:888–889
55. Van Voorhoeve JJC (1974) Organic fertilizers: problems and potential for developing countries. World Bank Fertilizer Background Paper No. 4. Office of the Economic Advisor, Washington, DC
56. Vlek PLG, Byrnes BH and Craswell ET (1980) Effect of urea placement on leaching losses of nitrogen from flooded rice soils. Plant Soil 54:441–449
57. Vlek PLG and Craswell ET (1981) Ammonia volatilization from flooded soils. Fert Res 2:247–259

58. Vlek PLG and Fillery IRP (1984) Improving nitrogen efficiency in wetland rice soils. Proc. No. 230. The Fertilizer Society, London, UK
59. Vlek PLG, Hong CW and Youngdahl LJ (1979) An analysis of N nutrition on yield and yield components for the improvement of rice fertilization in Korea. Agron J 71:829–833

218. Fluch, J., and H. J. P. Cleef, Ergebnisse eines . . . [faded text] . . . scannt the . . .
     [illegible] . . . B.V., [illegible] Buchler, [illegible] . . . [illegible]

219. Müller, Hans, Warm Acquisitions 1970 . . . [illegible] . . . [illegible] . . . 1970, [illegible] . . .
     and Wild companion, Bolina, a . . . [illegible] . . . of microbiological . . . Nova Acta . . .
     21 [illegible].

8. New developments in nitrogen fertilizers for rice

LJ YOUNGDAHL, MS LUPIN,[1] and ET CRASWELL[2]

International Fertilizer Development Center (IFDC), P.O. Box 2040, Muscle Shoals, Alabama 35662, USA

Key words: Nitrogen fertilizers, rice, fertilizer efficiency

Abstract. The efficiency of nitrogen (N) fertilizer products and practices currently used on rice is low, and improving this efficiency would be very beneficial to rice-growing countries. The development of new N fertilizers is best achieved by following a logical sequence of testing and evaluation procedures in a variety of settings from the laboratory to the farmer's field. Novel N fertilizers currently at various stages of testing include urea supergranules for deep placement, urea coated with various materials to control the N release rate, mixtures of a urease inhibitor with urea to reduce losses, and organic N sources other than urea.

New developments in nitrogen fertilizers for rice

Rice is very responsive to nitrogen (N) fertilization, and the high-yield potential of modern varieties cannot be realized without adequate N supply to the plant during the entire growing season. Over the last two decades, many rice-growing countries in Asia have therefore rapidly expanded their production and/or importation of N fertilizers, particularly urea [21]. This policy had paid dividends in increased rice production. However, rice is not very efficient at recovering fertilizer N in flooded field systems. Mitsui [20] quotes figures of 30–40% for use of broadcast fertilizer N by rice. Later work as part of a program of the International Fertilizer Development Center (IFDC) has confirmed these low figures. Craswell and Vlek [4] conducted a survey of data on fertilizer efficiency using the isotope ^{15}N. In this survey the efficiency of basal applications ranged from 7% to 38%, and recoveries from split applications ranged from 35% to 61%. In each case fertilizer is normally applied by broadcasting. Thus, the scope for improvement in fertilizer efficiency in rice is very great. In the tropical rice-growing countries of Asia the possible payoffs from a substantial increase in N fertilizer

[1] Present address: Dead Sea Works, Ltd., Beer-Sheva, Israel.
[2] Present address: Australian Centre for International Agricultural Research, Canberra, A.C.T., Australia.

149

Fertilizer Research 9 (1986) 149–160
© Martinus Nijhoff/Dr W. Junk Publishers, Dordrecht – Printed in the Netherlands

150

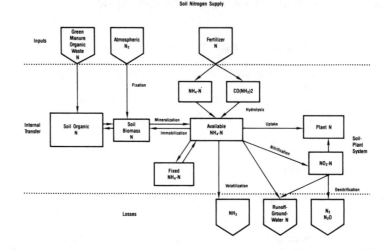

Figure 1. Nitrogen transformations in flooded soil systems (Source: Craswell and Vlek, 1982)

efficiency include increased net farmer income, reduced importation of foreign-produced fertilizers, and increased food supplies, which might possibly reduce the need to import food [21].

As a rule, an understanding of the fate of N applied to soil-plant systems is essential if rapid improvements in efficiency of use of N by crops are expected. Nitrogen can be lost by nitrification-denitrification, ammonia volatilization, leaching, and runoff in wetland in rice fields (Figure 1). The importance of each of these mechanisms depends on the form and the method of fertilizer application, the soil properties, climate, and the prevailing floodwater regime.

Recent research in the Philippines in lowland rice fields characterized by low percolation rates has highlighted the importance of ammonia volatilization [11]. This and earlier research [19] showed that the amount of NH_3 loss was markedly influenced by the management practice used. For example, the application of urea to the floodwater, a common farmers' practice, resulted in the loss of 50–60% of the N applied within 10 days. Surface incorporation of urea using harrows reduced the N loss by one-half as compared with the farmers' practice. Other studies have shown that NH_3 volatilization and total N loss are minimal when urea is deep placed in puddled soil [7, 19].

Numerous other studies on the fate of N applied to puddled rice soils have also illustrated the need for practices and materials that minimize the buildup of ammoniacal N in floodwater [3, 11, 30]. Different requirements are needed in soil-plant systems characterized by high percolation rates. In this case deep placement of urea-N, in particular, can cause high leaching losses

[15]. Such findings serve to illustrate the need for the development and evaluation of a range of practices and/or materials to meet the foal of improved N use in rice.

In addition to reducing N losses and increasing the uptake of N by the rice crop, the improved fertilizer products or practices should also be designed to maximize the grain yield per unit of fertilizer N absorbed. In this regard, the timing of N availability throughout the life of the crop may significantly influence fertilizer efficiency [8, 29]. Research with slow-release fertilizers in Korea has shown that the N supply during the period between 2 weeks after transplanting and maximum tillering critically influenced yield in rice [34].

This paper reviews progress in research to develop more efficient N fertilizers for use in lowland rice production in Asia.

Fertilizer evaluation

Experimental fertilizer materials are first evaluated in the laboratories and greenhouses of the International Fertilizer Development Center. Laboratory tests usually center on product quality (chemical analysis and size distribution) and, if appropriate, the nutrient release pattern. Techniques have been developed for measuring nutrient release under simulated lowland soil conditions [24]. This evaluation is followed by greenhouse trials. Pots containing 8 kg of puddled soil are fertilized with the experimental fertilizers and transplanted with a modern high-yielding rice variety. Details of the experimental methods were published earlier [5]. Nitrogen uptake and grain yield are measured to determine fertilizer performance in relation to a standard fertilizer treatment-split application of urea. Additionally, ^{15}N-labeled fertilizers are often used to provide additional information about the fate of the fertilizer N in the plant-soil system. The methods developed for these ^{15}N studies have been described in detail elsewhere [1].

The most promising fertilizer materials go on to field testing at other international centers, experiment stations, and evaluation networks. The most extensive network for N fertilizer evaluations in rice is the International Network on Soil Fertility and Fertilizer Evaluation for Rice (INSFFER). This network is a collaboration among IRRI, IFDC, and national research programs located primarily in Asia. This network functions in several ways to enhance national research activities. A network such as this allows the evaluation of a fertilizer over a wide range of environmental and cultural practices, provides national scientists with the latest advances in fertilizer technology, and promotes the exchange of information about soil fertility-related problems among national scientists.

Experimental fertilizer materials

Several concepts for improving fertilizer products have been developed. These concepts are largely based on the need to minimize the accumulation of ammonia in the floodwater which leads to volatilization, the most serious loss mechanism as shown by recent research [11, 30, 31]. The concepts developed are (1) slow release, (2) deep placement, and (3) urease inhibition. These concepts have been incorporated into a variety of experimental fertilizers which have been evaluated. Several design criteria have been set for the experimental fertilizers. These include (1) reduction of N losses, (2) proper timing of availability for the crop and climatic condition, (3) a minimum of changes in cultural practices, (4) low cost, (5) ease of handling, (6) high analysis, and (7) environmental safety.

Table 1. Network evaluation of sulfur-coated urea (SCU) relative to broadcast urea (BU) rice

| Program | Number of experiments | | |
	SCU > BU	SCU = BU	SCU < BU
INSFFER	50	51	2
INPUTS[a]	6	35	2
HFC[b]	29	36	6
Total	85	122	10
Percentage	39	56	5

[a] INPUTS = Increasing Productivity Under Tight Supplies
[b] HFC = Hindustan Fertilizer Corporation, Ltd

Slow release. The coating of urea granules with a material of low water solubility is the primary means used to delay the dissolution of readily soluble fertilizers. Sulfur-coated urea (SCU) is probably the most widely known coated urea material. It was developed by the Tennessee Valley Authority many years ago and has been extensively tested throughout the world. Table 1 shows how SCU compares with a broadcast urea application in several different series of field trials. The SCU was superior or equal to a broadcast application of urea in 95% of the farm-level trials. This performance is quite impressive since a single application of SCU is being compared with a split application of urea, and the single application results in a considerable saving in labor. Furthermore, SCU provides an input of sulfur which may be deficient in some areas [10]. However, the cost of producing SCU has limited its development into a commercial product. Many countries of the world do not have sulfur deposits and would have to import sulfur. The concept of coating urea to improve its efficiency is a good one, and many coatings other than sulfur are possible. One such coating, a sodium silicate base with a polymer final coat (PCU), has proven to be one of the best coating materials tested [10]. The urea release from these granules occurs in three phases. In the first 'delayed' phase little N is released. In the second 'maximum rate' phase most of the N is released ($\cong 70\%$), and in the last 'slow rate' phase the remaining fertilizer N is slowly released [26].

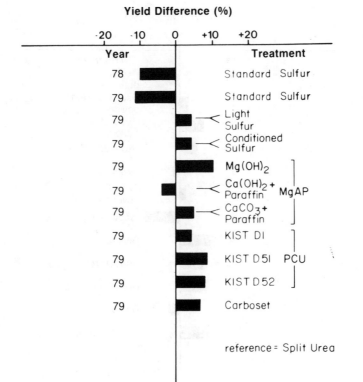

Figure 2. Percentage grain yield increase or decrease relative to a split application of urea for experimental coated products in greenhouse experiments

The rate of release can be modified by changing the chemical composition of the coating and the way in which the fertilizer is applied. In greenhouse pot trials this material (PCU, Figure 2) in various formulations has performed quite well compared with the standard split-urea applications.

Urea with magnesium ammonium phosphate-based coatings (MgAP, Figure 2) has also performed well compared with split applications of urea. Additionally, IFDC is cooperating with the Rubber Research Institute of Malaysia and Petroliam Nasional Berhad (PETRONAS) in reasearch on using rubber coatings, but problems exist with developing the technology for applying these coatings to urea at low cost. These nonsulfur coatings are still in the experimental stages and have not been extensively tested in the field.

In addition to coatings, slow release can be obtained through the use of organic compounds that contain N and are much less soluble in water than is urea. Many of these compounds have been tested by researchers in the past 20–30 years [22], and IFDC has surveyed some of these N compounds in

154

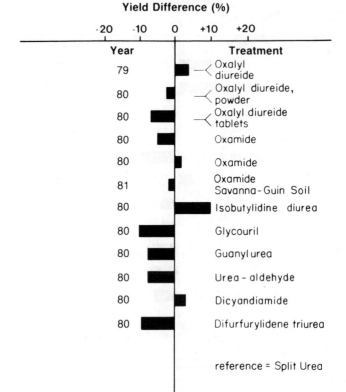

Figure 3. Percentage grain yield increase or decrease relative to a split application for broadcast and incorporated experimental organic N sources in greenhouse experiments

greenhouse trials (Figure 3). Most of the compounds do not perform very well compared with split urea, probably because the release is too slow for the growth cycle of the short duration high-yielding rice variety that was used. These compounds would be expected to perform better with longer duration crops including longer duration rice crops. The major problem of bringing organic N sources into commercial production is their cost. The production cost of most of these N compounds is considerably more than that of urea, and thus large increases in N efficiency would be necessary to justify their use. Research at IFDC is being continued with organic N sources, particularly in simple modifications of urea to reduce its solubility without appreciably increasing the cost of the product.

Deep placement. It has been found that placing fertilizer deep in the soil in a lowland field system greatly reduces losses and increases grain yields [4, 6, 7, 9]. However, placing prilled or granular urea deep in flooded soils poses a major problem. Urea prills are very hygroscopic, tend to cake, and often will not flow freely through a fertilizer applicator. To overcome this problem very large granules were developed, commonly referred to as urea super-granules (USG), which can be placed deep in the soil by hand or with a mechanical applicator. Large particles of urea can be manufactured either through granulation or through briquetting [17]. Experiments have shown that there is no agronomic difference between these two products. The production of USG through granulation is by necessity a large-scale factory process. Briquetting, the physical compaction of urea into large particles by pressure, can be either a factory or village-level process. Small briquetting machines are available and are currently in use in China with ammonium carbonate, a common N fertilizer in that country. Although the exact cost for the production of USG has not been determined, it has been estimated that factory-level production using granulation would cost 5–10% more than prilled urea [14, 27]. At this premium over prilled urea, the economic analysis of some INSFFER trials showed USG to be quite attractive. The combined response over the 17 sites indicates that farmers would have to use 29–31% less N as USG than as prills to achieve the same yield increase [13].

Ammonia volatilization losses are very small if little fertilizer N is dissolved in the floodwater after fertilizer placement [19]. The amount of N entering the floodwater when supergranules are used is dependent on the soil structure and the device used for fertilizer placement. Khan [16] evaluated nine deep-placement methods, four of which utilized USG. Of these four the total floodwater N 24 hours after application ranged from 6 to 15 ppm. The effectiveness of application methods seems to be primarily related to the closing of the soil opening after the USG is placed in the soil. More research is needed in the development of applicators for deep placement where placement by hand is not feasible.

Figure 4 shows the results of greenhouse trials comparing the grain yield of urea supergranules with that of broadcast incorporated urea. The deep-placed urea supergranules always outperformed the split area application as long as the granules were not placed too close to the plant, as in the 1977 data when high ammonia concentrations severly limited plant growth. Urea supergranules have also been tested extensively in farmers' fields. Table 2 shows results from many of these trials. Analysis of these field trials revealed that USG was generally superior to broadcast urea except in soils with high percolation rates. Urea supergranules are more susceptaible to leaching losses than is prilled urea that is uniformly distributed on the soil. Pot studies have shown that at a percolation rate of 8–9 mm/day 37% of the fertilizer N was lost by leaching from the USG, whereas only negligible amounts were lost from broadcast prilled urea [33]. This susceptibility to leaching losses

156

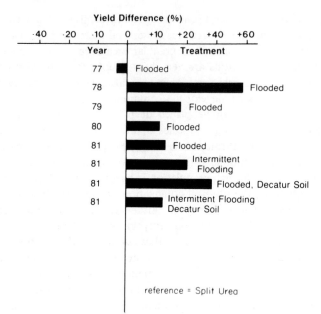

Figure 4. Percentage grain yield increase or decrease relative to a split application for deep placed urea supergranules in greenhouse experiments

has been corroborated by field results on light-textured soils in the Punjab area of India where USG performance was quite poor [15].

An additional advantage to the deep placement of urea, however, is that this mode of fertilization does not seem to interfere with dinitrogen fixation in the floodwater by blue-green algae [23]. Mineral N in the floodwater inhibits N_2 fixation and promotes the growth of green algae, which are not able to fix atmospheric N. Deep placement of urea does not allow much mineral N to enter the floodwater and thus allows the natural algal N_2-fixing system to continue providing N to the soil-crop system.

Table 2. Network evaluation of urea supergranules (USG) relative to broadcast urea (BU) for rice

Program	Number of experiments		
	USG > BU	USG = BU	USG < BU
INSFFER	38	65	3
INPUTS	18	31	0
HFC	39	27	5
Total	95	123	8
Percentage	42	54	4

[a]INPUTS = Increasing Productivity Under Tight Supplies
[b]HFC = Hindustan Fertilizer Corporation, Ltd

Inhibitors. When fertilizer N is placed in the soil it undergoes several transformations [25]. Of particular interest are the urease-catalyzed hydrolysis
of urea to ammonium and the biolgoical nitrification of ammonium to
nitrate (Figure 1). Inhibitors are chemical compounds that can be added
to the fertilizer to block these transformations. Researchers at IFDC have
had little or no success with nitrification inhibitors even in experiments in
which the soil is subjected to wetting and drying to encourage sequential
nitrification and denitrification [12]. Use of puddled soil to grow transplanted rice may have been responsible for this ineffectiveness of the nitrification inhibitors because it has been shown that puddling inhibits nitrate
accumulation in drying soils [28].

In contrast to the nitrification inhibitors, urease inhibitors offer great
promise for reducing N losses. One compound in particular, phenyl phosphorodiamidate (PPD), has been identified as a very effective inhibitor [35].
At IFDC, the PPD was incorporated with urea as a fine powder and the
mixture converted to granular form by the compaction process [18].
Figure 5 shows the effects of the timing of fertilizer applications with and
without PPD on fertilizer losses in greenhouse trials. The PPD always reduced
fertilizer losses, regardless of the timing of application, although the relative
effectiveness varied. It is thought that PPD slows the transformation of urea
to ammonium N, thus maintaining a lower ammoniacal-N concentration in
the floodwater. This slow hydrolysis of urea allows the fertilizer N to move
into the soil where it can be absorbed or immobilized, and this reduces the
ammonia volatilization losses. The presence of PPD delays the disappearance

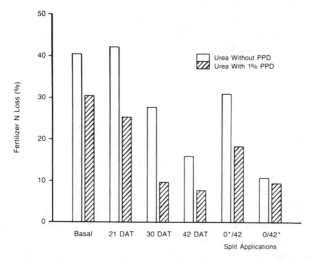

Figure 5. Fertilizer losses from ^{15}N-labeled urea with and without PPD at various times
of broadcast applications in a greenhouse experiment. (DAT = days after transplanting.
* = ^{15}N-labeled urea in split applications)

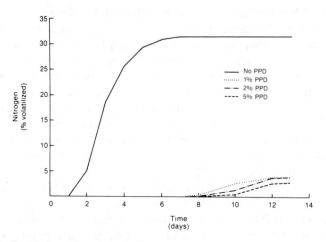

Figure 6. Cumulative ammonia volatilization losses from urea dn urea with PPD in a flooded soil system (Source: Byrnes, et al. 1983)

of urea in the floodwater from 3 days to 7 days. It also greatly reduces the ammoniacal-N concentration, which reached nearly 60 ppm after 2 days when no PPD was present. With PPD, the maximum ammonia level in the flood-water was only 20 ppm 5 days after fertilization. As can be seen in Figure 6, this change in the floodwater chemistry reduces ammonia volatilization from approximately 30% to 5% of the fertilizer N applied [2]. Research in this area is being continued to identify new compounds that have a longer lasting effect than does PPD. Also, compounds that are lower in cost are being sought. Urease inhibitors hold the promise of increasing fertilizer efficiency without requiring major modifications in fertilizer production facilities or in the cultural practices of the small farmer.

Conclusions

Several strategies exist for improving N fertilizer efficiency through modifi-cations of the fertilizer product. Of these, deep placement is the closest to farm-level acceptance. The manufacture of large granules is not difficult, and the product could be available to small farmers in the relatively near future. In the longer term, the inclusion of urease inhibitors in urea may provide the greatest increases in N fertilizer efficiency. Research is also being done on coatings to explore the use of indigenous materials such as plant waxes from sugarcane or rice bran and also to improve mineral coatings such as sulfur. Although much remains to be discovered on how to use these materials, they do offer some potential. These and other fertilizer technologies are expected to be developed in the next few years. Opportunities exist for improving fertilizer efficiency in most countries, and cooperation between the international research centers and national agricultural research organizations is speeding the process of transferring fertilizer technology and know-how to the small farmers.

References

1. Buresh RJ, Austin ER and Craswell ET (1982) Analytical methods in ^{15}N research. Fert Res 3:37–62
2. Byrnes BH, Savant NK and Craswell ET (1983) Effect o a urease inhibitor phenyl phosphorodiamidate on the efficiency of urea applied to rice. Soil Sci Soc Am J 47:270–274
3. Craswell ET, De Datta SK, Obcemea WN and Hartantyo M (1981) Time and mode of nitrogen fertilizer application to tropical wetland rice. Fert Res 2:247–259
4. Craswell ET and Vlek PLG (1979) Fate of fertilizer nitrogen applied to wetland rice. In Nitrogen and Rice, pp. 175–192, IRRI, Los Baños, Philippines
5. Craswell ET and Vlek PLG (1979b) Greenhouse evaluation of nitrogen fertilizers for rice. Soil Sci Soc Am J 44:1184–1188
6. Craswell ET and Vlek PLG (1982) Nitrogen management for submerged rice soils. In Proceedings 12th International Congress of Soil Science, New Delhi, 3:158–181
7. De Datta SK, Fillery IRP and Craswell ET (1983) Results from recent studies on nitrogen fertilizer efficiency in wetland rice. Outlook on Agriculture 12:125–134
8. De Datta, SK, Magnaye CP and Magbanua JT (1969) Response of rice varieties to time of nitrogen application in the tropics. In Proceedings Symp. (Trop Agric Res) pp. 73–87, Optimization of fertilizer effects in rice, Tokyo, Japan
9. De Datta SK, Magnaye CP and Moomaw JC (1968) Efficiency of fertilizer nitrogen (N^{15}-labelled) for flooded rice. Trans. 9th Int. Soil Sci. Congr., Adelaide, Australia, 4:67–76
10. De Datta SK, Stangel PJ and Craswell ET (1981) Evaluation of nitrogen fertility and increasing fertilizer efficiency in wetland rice soils. In Proceedings Symp. on Paddy Soils, pp. 171–206, Science Press, Beijing, People's Republic of China
11. Fillery IRP, Simpson JR and De Datta SK (1984) Influence of field environment and fertilizer management on ammonia loss from flooded rice. Soil Sci Soc Am J 48:914–920
12. Fillery IRP and Vlek PLG (1982) The significance of denitrification of applied nitrogen in fallow and cropped rice soils under different flooding regimes. Plant and Soil, 65:153–169
13. Flinn JC, Mamaril CP, Velasco LE and Kaiser K (1984) Efficiency of modified urea fertilizers for tropical irrigated rice. Fert Res 5:157–174
14. Holte E, Ryan J and Ostlyngen TW (1982) Pan granulation process for urea super-granules. Paper presented at the International Workshop/Training Course on Nitrogen Management, Fuzhou, China
15. Katyal JC, Singh BJ, Sharma VK and Craswell ET (1985) Efficiency of some modified urea fertilizers for wetland rice grown on a permeable soil. Fert Res (submitted)
16. Khan AV (1983) Deep placement fertilizer applicators for wet paddies. Paper presented at the 1983 winter meeting of the American Society of Agricultural Engineers, Chicago, Illinois
17. Lupin MS, Lazo JR, Le ND and Little AF (1983) Briquetting. Tech. Bull. T-26, International Fertilizer Development Center, Muscle Shoals, Alabama
18. Lupin MS and Le ND (1983) Compaction. Tech. Bull. T-25, International Fertilizer Development Center, Muscle Shoals, Alabama
19. Mikkelsen DS, De Datta SK and Obcemea WN (1978) Ammonia volatilization losses from flooded rice soils. Soil Sci Soc Am J 42:725–730
20. Mitsui S (1954) Inorganic nutrition, fertilization, and soil amelioration for lowland rice. Tokyo: Yokendo Ltd
21. Mudahar MS and Hignett TP (1983) Energy and fertilizer. Tech. Bull. T-20, International Fertilizer Development Center, Muscle Shoals, Alabama
22. Murray TP and Horn RC (1979) Organic nitrogen compounds for use as fertilizers.

Tech. Bull. T-14, International Fertilizer Development Center, Muscle Shoals, Alabama

23. Roger PA, Kulasooriya SA, Tirol AC and Craswell ET (1980) Deep placement: A method of nitrogen fertilizer application compatible with algal nitrogen fixation in wetland rice soils. Plant and Soil 57:137–142

24. Savant NK, Clemmons JR and James AF (1982) A technique for predicting urea release from coated urea in wetland soil. Commun. Soil Sci Plant Anal 13:793–802

25. Savant NK and De Datta SK (1982) Nitrogen transformation in wetland rice soils. Adv Agron 35:241–302

26. Savant SK, James AF and McClellan GH (1983) Urea release from silicate- and polymer-coated urea in water and a simulated wetland soil. Fert Res 4:191–199

27. Schultz JJ, Polo JR and Le ND (1984) Technical/economic assessment of producing and marketing urea supergranules (USG) in Indonesia. Paper presented at the Urea Deep Placement Workshop, Bogor, Indonesia

28. Sompongse D, Craswell ET and Byrnes BH (1983) Effect of soil puddling on loss of ^{15}N-labeled urea under various flooded conditions. Agron Abs 1983, Amer Soc Agron, Madison, Wisconsin

29. Tanaka A, Patnaik S and Abichandani CT (1959) Studies on the nutrition of rice plant (*Oryza sativa* L.) III. partial efficiency of nitrogen absorbed by rice plant at different stages of growth in relation to yield of rice (*O. sativa* var. *indica*). Proc Ind Ac Sci 49:207–216

30. Vlek PLG and Craswell ET (1979) Effect of nitrogen source and management on ammonia volatilization losses from flooded rice soil systems. Soil Sci Soc Am J 43:352–358

31. Vlek PLG and Craswell ET (1981) Ammonia volatilization from flooded soils. Fert Res 2:227–245

32. Vlek PLG and Stumpe JM (1978) Effect of solution chemistry and environment conditions on ammonia volatilization losses from aqueous systems. Soil Sci Soc Am J 42:416–421

33. Vlek PLG, Byrnes BH and Craswell ET (1980) Effect of urea placement on leaching losses of nitrogen from flooded rice soils. Plant Soil 54:441–449

34. Vlek PLG, Hong CW and Youngdahl LJ (1979) An analysis of N nutrition on yield and yield components for the improvement of rice fertilization in Korea. Agron J 71:829–833

35. Vlek PLG, Stumpe JM and Byrnes BH (1980) Urease activity and inhibition in flooded soil systems. Fert Res 1:191–202

9. Improving nitrogen fertilization in mechanized rice culture

DM BRANDON and BR WELLS[1]

Rice Research Station, Louisiana Agric. Exp. Stn., Louisiana State Univ. Agric. Center and Dep. of Agronomy, Univ. of Arkansas, respectively

Key words: soil submergence, N efficiency, slow release N, plant analysis, cultural system, water-seeded, dry-seeded

Abstract. Rice is produced in highly mechanized and energy intensive water-seeded and dry-seeded systems in the United States. Nitrogen fertilization management relative to N source and time of application differs in the two systems because of the timing of soil submergence which influences N retention in the soil. Nitrogen management studies show that N fertilizer efficiency is maximized in water-seeded rice when ammonical N is placed 5 to 10 cm in the soil immediately before flooding. Nitrogen applied on a dry soil surface immediately before flooding dry-seeded rice results in N movement into the soil and retention for plant utilization. Nitrogen application preplant or into water after flooding results in N losses in dry-seeded rice. Split N application gives acceptable N efficiency when 65 to 75% of the total N fertilizer requirement is applied preflood followed by a midseason N topdressing. Sulfur-coated urea and nitrapyrin soil incorporated with urea reduce N loss in dry-seeded rice. Total N requirements of rice in the cultural systems is dependent on cultivar, soil N fertility and other factors. Plant analysis research establishes critical N concentrations in semidwarf and tall rice cultivars in the water-seeded system.

United States rice production is highly mechanized and energy intensive. The total labor required to produce a crop is only 11 to $13\,h\,ha^{-1}$ [6, 9]. Mechanized rice culture requires large energy inputs because irrigation, land preparation, chemical applications, and heated air drying of grain utilize power from internal combustion or electrical sources [8, 9].

Rice is produced in two distinctly different cultural systems that require different N fertilization management for optimizing N efficiency. A dry-seeded, delayed flood cultural system is used predominantly in Arkansas, Mississippi, Missouri, and Texas. A water-seeded, permanently flooded system is used in California and Louisiana. The objective of this paper is to identify the time and method of N fertilization for optimum N efficiency in the two mechanized cultural systems.

[1] Professor Rice Res. Stn., P.O. Box 1429, Crowley, LA, 70527-1429, and Dep. of Agronomy, Univ. of Arkansas, Fayetteville, AR, 72701, respectively.

Fertilizer Research 9 (1986) 161−170
© Martinus Nijhoff/Dr W. Junk Publishers, Dordrecht − Printed in the Netherlands

Description of mechanized rice cultural systems

Water-seeded permanently flooded system

The seedbed for water-seeded rice is prepared by dry tillage to provide a slightly cloddy, or grooved soil surface free of large clods, raised areas and depressions. Levees are surveyed at 3 to 16 cm elevation differentials to permit rapid flooding and drainage, and maintenance of a uniformly shallow water depth of 5 to 10 cm during the growth period. Presoaked seed is aerially broadcast into the flooded field which is flooded continuously or is temporarily drained to promote seedling anchorage to the soil [8]. Soil moisture is not allowed to dry below saturation after water-seeding even though it may be temporarily drained to enhance stand establishment. Rice seedlings emerge through the flood water 10 to 15 days after seeding depending on seedling vigor, water temperature, water depth, and other factors. Pesticides and fertilizers applied after flooding are distributed into the flooded field by airplane.

Dry-seeded delayed flood system

Dry seeded rice cultural systems involve land and seedbed preparation similar to the water-seeded system. The finished seedbed for dry-seeded rice, however, should be smooth and free of a cloddy surface. Rice is drill- or broadcast-seeded and incorporated into a moist seedbed. Depth of seed placement is critical, especially with semidwarf varieties and should be 1.0 to 4.0 cm [3]. Seedling emergence requires 8 to 20 days depending on seedling vigor, soil moisture, temperature, planting depth, and other factors. Dry-seeded rice is grown under upland conditions until seedlings are in the 4th to 5th leaf stage, although earlier flushing with irrigation water may be required to prevent moisture stress. A shallow flood of 5 to 10 cm is established when the rice seedlings are in the 4 to 5 leaf stage and it is maintained until grain maturity.

Nitrogen fertilization in mechanized culture

Nitrogen fertilization management relative to N source, timing and method of application differs between the two mechanized rice cultural systems. Ammonical N sources are preferable to nitrate sources because NH_4-ions can be retained in a submerged soil when placed into the reduced zone. The NO_3-ion cannot be retained in a submerged soil because biological reduction of the NO_3-ion results in N loss from the soil [7, 10]. The relative efficiency of N fertilizer in both water-seeded and dry-seeded systems depends largely on time of N application in relation to flooding rice permanently, and development of the rice plant. Water management after N fertilizer application determines how much N is retained in the soil for plant utilization in both cultural systems.

Water-seeded cultural system

The time and rate of N required for optimum grain yield is dependent on native soil N, rice cultivar and other factors [1, 2]. Ammonical N placed 5 to 10 cm in the soil within 5 days before flooding results in good N retention in the soil if the soil is submerged until grain maturity [7, 10]. Alternate flooding and draining of the soil after N application results in nitrification-denitrification losses that greatly reduce N efficiency. Preplant application of a 90 kg N ha^{-1} suboptimum N rate results in greater rice yields than split applications of an equal N rate in the water-seeded system (Figure 1). Split N applications of 90 kg N ha^{-1} results in early plant N deficiency which reduces the number of panicles and grain yields (Table 1). The depression of grain yield due to split N application is not only caused by early plant N deficiency which reduces the number of panicles, but it is also caused by reduced N efficiency due to surface broadcasting N into the flooded system [4]. Split applications of N at the excessive 180 kg N ha^{-1} rate result in a more optimum N concentration in the plant than a single preplant 180 kg N ha^{-1} rate because the excessive preplant N causes severe crop lodging.

Fertilizer N requirements of rice usually depend on soil N fertility, rice cultivar and other factors. The optimum N fertilizer rate in a N fertile soil (Figure 2) is usually less than that required in N deficient soils (Figure 3). Semidwarf cultivars such as Bellemont (BLMT), Leah, and Lemont (LMNT)

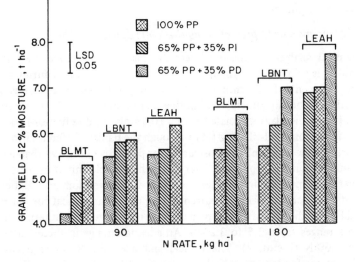

Figure 1. Effect of a single preplant (PP) N application and split N applications of 65 percent N PP plus 35 percent N topdressed at either the panicle initiation (PI) or panicle differentiation (PD) growth stages on yields of water-seeded rice

164

Table 1. Effect of N fertilizer rate and time of application on the grain yield and yield components of the semidwarf M9 cultivar

Total N	Time of N*				Panicles m^{-2}	Spikelets $panicle^{-1}$	100-Grain weight	Grain yield
	PP	MT	PI	FL				
	kg ha^{-1}						g	t ha^{-1}
0	–	–	–	–	367	24	2.85	3.5
101	101	–	–	–	613	42	2.75	7.5
101	67	34	–	–	563	39	2.83	6.8
101	67	–	34	–	581	42	2.85	7.2
101	67	–	–	34	566	38	2.88	7.0
135	135	–	–	–	689	42	2.73	8.5
168	168	–	–	–	639	48	2.60	7.2
168	112	56	–	–	608	44	2.78	7.8
168	112	–	56	–	601	45	2.65	7.6
168	112	–	–	56	636	45	2.80	7.8
LSD$_{0.05}$					82	7	0.13	0.9
C.V., %					9.9	12.1	3	10

*Time of N: PP = preplant; MT = midtillering; PI = panicle initiation; FL = flag leaf

require 20 to 40 kg N ha^{-1} more than the tall Lebonnet (LBNT) cultivar but the U.S. semidwarf cultivars also perform well as slightly suboptimum to excessive N rates (Figures 2, 3). Management of the overall rice production system so that all agronomic inputs are properly timed in relation to plant growth and development results in optimum N efficiency. Management practices that slow rice growth and development such as delayed flooding, fertilization, and pest control usually decrease N fertilizer efficiency and increase N requirement.

Use of plant analysis to predict N requirements of Rice

Chemical analysis of the most recently matured rice leaves (Y-leaf) for Kjeldahl N is valuable in estimating the N requirement of rice during growth and development of the crop [1, 5]. Use of plant analysis in rice is based conceptionally on the principle of limiting factors and critical N concentration (level of N in the plant associated with a 10% grain yield reduction). Multiple leaf samples from mid-tillering (MT) through the panicle differentiation (PD) stage provide more complete information for predictive purposes than a single leaf sample. Samples collected later than the panicle initiation (PI) stage, however, may provide information too late for prevention of N deficiency induced yield reduction in the current crop [1, 5]. The critical concentration of total N in the Y-leaf at the panicle initiation stage of southern U.S. rice cultivars ranges from 2.5 to 3.2% N. An adequate range in these cultivars is 2.6 to 3.6% N (Figure 4). The critical and adequate N concentration are usually greater for long-grain and semidwarf cultivars than for medium-grain cultivars [1, 5]. When the plant N concentration at panicle initiation is at or below the critical N concentration, an application of 30 to 40 kg N ha^{-1}

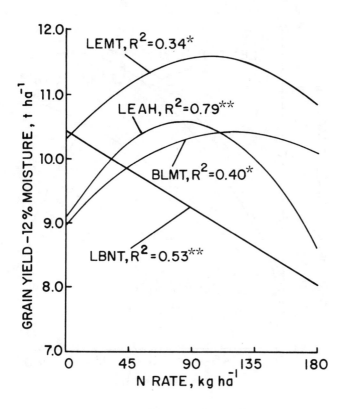

Figure 2. Grain yield response of semidwarf (Bellemont, Leah and Lemont) and tall (Lebonnet) rice cultivars to N fertilizer in a N fertile soil

will increase grain yield or prevent further grain yield reduction. Loss of grain yield potential because of early N deficiency or poor management practices is not completely reversible by N fertilization later in the season. Grain yield determinants such as panicles per m^2 and spikelets per panicle are determined relatively early in the life of the rice plant. Delayed N fertilizer application when inadequate preplant N is applied often results in yield loss because of reduction in the number of productive tillers (Table 1). Rice plant analysis for N is a valuable diagnostic aid to determine the N needs of rice during the growing period when used with experience and observation in the water-seeded cultural system.

Dry-seeded cultural system

Nitrogen applied preplant in the dry-seeded rice system is subject to nitrification because of aerobic soil conditions during the seeding to flooding (3 to 6 week) interval. Rapid and excessive loss of NO_3-ions due to denitrification occurs after flooding the soil when N fertilizer is applied preplant in

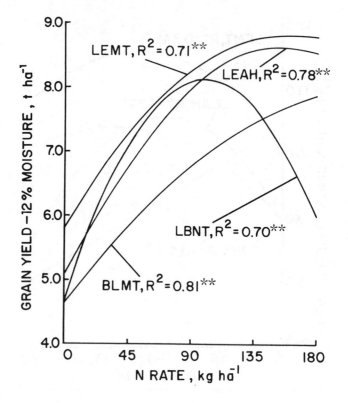

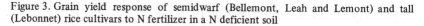

Figure 3. Grain yield response of semidwarf (Bellemont, Leah and Lemont) and tall (Lebonnet) rice cultivars to N fertilizer in a N deficient soil

the dry-seeded system. Therefore, nitrogen management in the dry-seeded system relative to time and method of N fertilization for maximum N efficiency differs from that of water-seeded rice.

Research conducted in Arkansas shows that rice reponds well to split-topdress applications of nitrogen [13] if the early season application is made onto dry soil just prior to flooding when the rice is at the 4 to 6 leaf growth stage. Figure 5 gives an indication of why urea N applied in this manner results in satisfactory uptake by the flooded rice plant. Urea applied to the dry soil moves downward with the wetting front of the floodwater to the 6 cm depth of a Crowley silt loam soil (Typic Albaqualf, fine, mont-morillonitic, thermic), thus moving it into the reduced soil zone. However, when the urea is applied into the floodwater most of the N remains in the 0–1 cm layer of soil where it is subject to possible loss by either volatilization [4] or denitrification [7]. Dry matter production of Starbonnet rice verified these results with linear increase of dry matter with increasing N rate when the urea was topdressed on the dry soil compared to no increase in dry

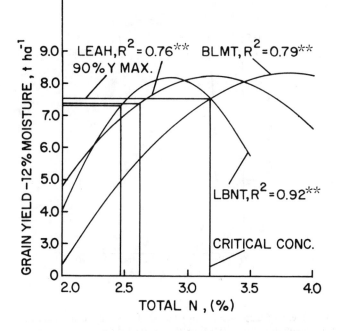

Figure 4. Estimated critical Y-leaf N concentration of semidwarf (Bellemont, Leah and Lemont) and tall (Lebonnet) rice cultivars at panicle initiation based on 10 percent yield reduction caused by N deficiency

matter production with increasing N rate when the urea was applied into the floodwater one day after flooding.

Nitrogen applied at the beginning of internode elongation may be applied in the irrigation water and still be effective for increasing grain yields. Grain yield data in Figure 6 shows 3 years averages for response of Starbonnet rice to both early season N as urea applied onto a dry soil surface and mid-season N applied into the floodwater. Apparently uptake by the rice plant at the early reproductive stage is sufficiently rapid to alleviate partially the problems of ammonia volatilization and denitrification normally associated with placing urea into the flood water. Research conducted several years ago utilizing ^{15}N tagged NH_4NO_3 indicated that most of the applied nitrogen entered the plant within a four day interval at this growth stage (B.R. Wells, 1960. M.S. Thesis, on file U of A Library).

Water management throughout the growing season has a major impact on grain yields of rice growing in N deficient soils. Wells [14] reports a 1.3 t/ha increase in grain yields of rice as an average across preplant N rates when flooding was continuous as compared to intermittent. The differential was 0.7 t/ha when the N was applied in split-topdress applications.

168

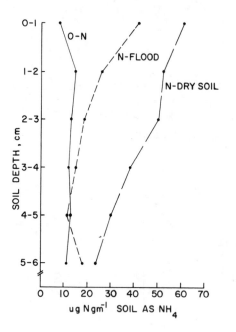

Figure 5. Movement of urea N into Crowley silt loam soil when broadcast on dry soil before flooding and into water after flooding dry-seeded rice

Use of nitrification inhibitors and controlled release N sources in dry-seeded rice culture

The potential NH_3 volatilization and denitrification losses of N associated with either preplant or split-topdress N applications in dry-seeded rice emphasize the need for a suitable preplant N fertilization system that will prevent nitrification during the interval of seeding to flooding. The recent escalation of aerial applications costs also gives an economic incentive to the search for a method of stabilizing the preplant N over this interval.

Nitrogen stabilization research in the southern U.S. has investigated the use of controlled release N sources and nitrification inhibitors to improve preplant N fertilizer efficiency. Research with various formulations of sulfur-coated urea (SCU) and isobutylidene diurea (IBDU) indicates that either of these sources applied preplant result in rice yields comparable to those achieved with split-topdress applications of urea or ammonium sulfate [11]. These products appear to stabilize N regardless of water management practices, but product cost limits their use in rice.

The efficacy of nitrification inhibitors in conserving preplant N also has been investigated in dry-seeded rice. Compounds such as nitrapyrin (2-chloro-6-(trichloromethyl)-pyridine have been soil incorporated preplant with urea [12] and anhydrous ammonia to delay nitrification, and subsequently,

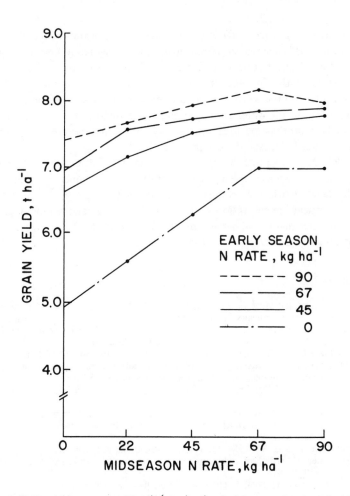

Figure 6. Grain yield response of Starbonnet rice to varying rates of early and midseason topdressed N in dry-seeded cultural systems

denitrification loss of N after flooding rice. These compounds have given N efficiency comparable to that obtained with split-topdress N application. The main limitations of materials such as nitrapyrin appear to be volatility, short lifetime in the soil, and sensitivity to soil texture, pH, organic matter, and temperature. The current nitrification inhibitors have had limited success in commercial rice production because of these limitations and the high level of management required for their use.

Summary and conclusions

Mechanized rice production in the United States requires specific fertilizer timing and methods of N application for maximum N efficiency and grain yields. Nitrogen fertilizer efficiency is maximized in water-seeded rice when ammonical N is placed 5 to 10 cm in the soil immediately before flooding. Broadcast application of urea or ammonical N on the soil surface immediately before permanently flooding dry-seeded rice results in movement of N into the soil profile by the wetting front and retention of N in the submerged soil. Broadcast N into the water after flooding dry-seeded rice is less efficient than preflood N. Sulfur-coated urea and nitrapyrin soil incorporated with urea at planting reduced N loss in dry-seeded rice.

Kjeldahl N in the most recently matured rice leaves is positively correlated with grain yield when sampled at panicle initiation and panicle differentiation growth stages. Critical N concentrations range from 2.5 to 3.6 percent N with the critical concentration dependent on cultivar and sampling growth stage. Nitrogen plant analysis is used to determine the N fertilizer requirement during the growth of the rice crop.

References

1. Brandon DM RP Mowers RH Brupbacher HF Morris WJ Leonards TR Laing SM Rawls and NJ Simoneaux (1982) Nitrogen requirements of new rice varieties and relationship between Y-leaf N and grain yield. p. 52–100. In 74th Annual Prog Rep Rice Res Stn, Louisiana State Univ, Crowley, Louisiana
2. Brandon DM HL Carnahan JN Rutger ST Tseng CW Johnson JF Willians CM Wick WM Canevari SC Scardaci and JE Hill (1981) California rice varieties: Description, performance and management. Univ California Spec Pub. 3271
3. McKenzie KS and DM Brandon (1982) Leah – a new rice variety. Louisiana Agric 25(4):8–10
4. Mikkelsen DS and SK DeDalla (1979) Ammonia volatilization from wetland rice soils. p. 135–156. In IRRI Nitrogen and Rice. Int Rice Res Int, Los Baños, Phillippines
5. Mikkelsen Duane S (1983) Diagnostic plant analysis for rice. p. 30–31. In Reisenauer HM (ed) Soil and plant tissue testing in California. Div Agric Sci, Univ California Bull 1879
6. Musick JA and ME Salassi (1983) Projected cost and return of rice and soybeans in southwest Louisiana. Dep Agric Econ, Louisiana State Univ Res Report 607
7. Patrick WH Jr and DS Mikkelsen (1971) Plant nutrient behavior in flooded soil. p. 187–215. In Olsen RA Army ATJ Hanway JJ Kilmer VJ (eds) Fertilizer technology and use. 2nd ed. Soil Sci Soc Am, Madison, Wisconsin
8. Rutger J Neil and D Marlin Brandon (1981) California rice culture. Scientific American 244:42–51
9. Scardaci SC (1983) Rice production costs in Colusa, Glenn and Yolo countries. Univ California Coop Ext Pub, Colusa, California
10. Tusneen ME and WH Patrick Jr (1971) Nitrogen transformation in waterlogged soil. Louisiana State Univ Bull 657
11. Wells BR and PA Shockley (1975) Conventional and controlled-release nitrogen sources for rice. Soil Sci Soc Amer Proc 39:549–551
12. Wells BR (1977) Nitrapyrin (2-chloro-6-(trichloromethyl)-pyridine) as a nitrification inhibitor for paddy rice. Down to Earth. 32:28–32
13. Wells BR and Wade F Faw (1978) Short-statured rice response to seeding and N rates. Agron J 70:447–480
14. Wells BR and PA Shockley (1978) Response of rice to varying flood regimes on a silt loam soil. El Riso. 27:81–87

10. Improving nitrogen fertilizer efficiency in lowland rice in tropical Asia

SK DE DATTA

International Rice Research Institute, P.O. Box 933, Manila, Philippines

Key words: improved water management, best split, mechanisms of N losses, varietal differences in N use efficiency, deep placement, nitrification and urease inhibitors

Abstract. Rice production in Asia must increase 2.2–2.8% annually to keep abreast of increasing population. Greater fertilizer use and crop intensification together with varietal improvement and investment in irrigation will all contribute to increased rice supply. Because fertilizer and input prices have risen faster than the price of rice, increasing fertilizer N efficiency will be a major challenge for rice researchers and farmers. Greater fertilizer N efficiency may be achieved through improved timing and application methods, and particularly through better incorporation of basal fertilizer N without standing water. Other promising alternative practices are use of N-efficient rice varieties, hand or machine deep placement of urea supergranules, and use of slow release N fertilizers. Research challenges include development of placement machines for prilled urea and identification of cost-efficient nitrification and urease inhibitors. Under the present resource-scarce situation in many tropical Asian countries, several complementary practices must be followed to supplement inorganic N sources. Fertilizer supplies and proper price support should be maintained and wherever possible increased, and appropriate fertilizer materials and application methods must be devised to increase N use efficiency in lowland rice, so that increasing rice requirements are fulfilled.

Introduction

Rice currently represents as much as 75% of the caloric intake of the 2 billion people living in Asia. It is the most important food crop in the region where population densities are highest and overall dietary levels are least adequate. In the vast monsoonal areas of tropical Asia, rice gives the highest food-staple cereal yield from a fixed land area of arable land.

In many tropical rice-growing areas, particularly in South and Southeast Asia, lowland rice culture predominates. Under lowland culture, land is either prepared wet by puddling the field or prepared dry, but water is always held in the field by bunds [6].

In Asia, rice production has increased an average 2.7% annually — slightly faster than population growth, but somewhat slower than the growth in demand [17]. Yield has increased because of the adoption of modern varieties and associated technology, increased year-round irrigation facilities, and more fertilizer use.

Fertilizer Research 9 (1986) 171–186
© *Martinus Nijhoff/Dr W. Junk Publishers, Dordrecht – Printed in the Netherlands*

These spectacular yield increases still do not reflect the potential that researchers believe the technology has at the farm level in tropical Asia. There is general agreement among policy makers and analysts that government options to increase rice production include a combination of:
1. improved water management,
2. increased productivity from non-irrigated areas,
3. varietal improvement,
4. increased use of fertilizer and pest control, and
5. appropriate agricultural policies and institutions.

As the need for higher rice yields and production becomes more urgent, and as prices of energy-related materials such as fertilizers constantly increase, efforts to increase fertilizer N efficiency in lowland rice should include detailed studies on fertilizer nutrient use and crop management practices.

My paper summarizes some of the current knowledge on management practices for improving N use efficiency in lowland rice soils.

Nitrogen requirements for rice

Rice plants require as much N as possible at early and mid-tillering to maximize panicle number. Plants also need N at reproductive and ripening stages to produce optimum spikelets per panicle and percentage filled spikelets.

N requirement by the rice crop can be ascertained by: visual symptoms of the crop, plant analyses, soil analyses, and yield response with fertilizer application. The last method is the surest, but the other methods can provide supplementary information.

Nitrogen use efficiency

Research on constraints to high yield in farmers' fields in the Philippines has shown that insufficient fertilizer or inappropriate fertilizer management accounts for one-half to two-thirds of the gap between, actual and potential rice yields in farmers' fields [7]. These yield differences were determined by comparing farmer's fertilizer practices with the best available technology; the *best-split* application.

Although the high N content and low price of urea make it the most popular fertilizer for tropical rice, it is not an ideal fertilizer for the lowland rice crop. No single urea fertilizer or urea fertilizer management practice is suited to all crop situations because of diversity of soil and climatic regimes under which rice is grown. Alternative strategies must be developed to increase fertilizer efficiency in tropical lowland rice.

Efficiency of fertilizer nitrogen utilization

The percentage of N recovery by rice varies with soil properties, methods, amount and timing of fertilizer application, and other management practices.

N fertilizer recovery usually is 30–50% in the tropics [24]. Percentage of N recovery tends to be higher at low N application levels and when N is placed deep in soil or topdressed at later growth stages.

N utilization efficiency for grain production in the tropics is about 50 kg rough rice/kg nitrogen absorbed, and is almost constant regardless of rice yield [32]. N efficiency is about 20% higher in temperate regions than in the tropics.

Using values for the recovery percentage and utilization efficiency obtained for the tropics, Yoshida [32] estimated fertilizer N efficiency to be 15–25 kg rice/kg applied N. These values were confirmed in N response experiments [24].

The primary goal of improved N management practices should be to maximize N uptake at critical growth stages and minimize transformation processes that lead to losses or temporary losses of N from soil water systems [12]. It also is critical to ensure that N absorbed by the plant is used for grain production, which Murayama [22] called *productive efficiency*. Productive efficiency is the amount of rice produced per kilogram of nutrient applied.

Poor utilization of N fertilizer by rice is thought to be largely due to N losses from the soil plant system. Ammonia volatilization, denitrification, runoff, and leaching are major mechanisms of loss [3]. Obviously, the nature of the N fertilizer, the application method, and field-level conditions all affect fertilizer efficiency [26].

Management practices for fertilizer nitrogen

Since modern high-yielding rice varieties were introduced in tropical Asia in the mid-1960's, researchers have sought to improve fertilizer N efficiency in flooded rice culture. The need for increased efficiency has been further emphasized by recurring energy shortages, a fertilizer N shortage, and rising prices. Furthermore, fertilizer subsidies have been discontinued or reduced in several countries (Figure 1), which make high rates of fertilizer use unlikely. Shortages, plus the heavy losses of fertilizer N in rice fields, have encouraged fertilizer technologists and agronomists to develop more efficient fertilizer materials and management practices.

Payoffs from investments in agricultural research are high in irrigated areas [31], but most rice farmers still depend on rainwater to sustain their crops.

High levels of inputs such as fertilizers are not used even in irrigated rice. For rainfed areas, the input levels are even less [29]. For example, constraints research in rainfed conditions in 1981–83 wet seasons showed that 57% of a 1.4 t/ha yield gap in 2 Philippine provinces is caused by low rates and poor timing of fertilizer N applications by 84 farmers (Figure 2).

Several alternative practices have been evaluated at IRRI and elsewhere to increase N use efficiency in lowland rice [11]. They are:

174

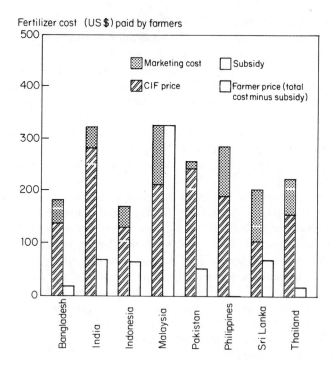

Figure 1. Total cost (CIF + marketing cost), fertilizer subsidy, and farmer price of fertilizer in selected Asian rice growing countries. From FAO [16]

1. varietal differences in N utilization efficiency;
2. improved timing of N application;
3. deep placement of N fertilizer;
4. controlled release N fertilizers;
5. use of nitrification and urease inhibitors; and
6. complementary practices to improve N use efficiency.

1. Varietal differences in nitrogen utilization efficiency

Successful exploitation of varietal differences in ability to utilize soil N is important because soil is the major N source in lowland rice. Results from 1976–1982 experiments at four stations in the Philippines suggest that IR42 uses soil and fertilizer N more efficiently than IR36, the most widely grown rice variety in the world [11]. Subsequent research showed that up to 30 days after transplanting (DT) medium duration IR42 uses soil N more efficiently than IR9729-67-3, a very early duration rice. After 30 DT, IR42 utilizes fertilizer N better than IR9729-67-3 (Figure 3).

In IRRI-University of California-Davis collaborative research, ^{15}N-depleted ammonium sulfate was used in 1981–82 dry and wet season field experiments to analyze grain yield and N uptake of several rices. Results suggested that

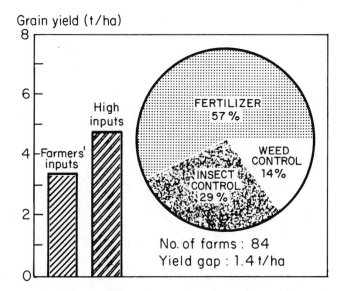

Figure 2. Average yields with farmers' inputs and high inputs and contributions of fertilizer, insect control, and weed control to the yield gap in rainfed farms in Tarlac, Libmanan, and San Fernando in Camarines Sur, Philippines, 1981–83 wet seasons

grain yield and total N uptake, and total quantity of N used is related to growth duration of rice (Table 1). The total quantity of N used by the rice plant is related to duration group. In 1981 dry season, the long duration rices absorbed more total N but less fertilizer N than shorter duration varieties. However, in 1981 wet season, total N uptake was highest in the medium duration group. In 1982 dry season, short duration rices absorbed less N than medium and long duration varieties, which absorbed similar amounts. In general, short duration rices used more fertilizer N. In 1983, an additional 78 rices were evaluated to determine consistent differences in N fertilizer response, with the objective of selecting rices that would perform well with minimum fertilizer N.

Results from 3-year trials showed significant differences in fertilizer N utilization based on growth duration, tillering, and possibly other plant characteristics. Preliminary results suggest that short season varieties depend more on supplementary fertilizer N in the dry season than long duration varieties [1].

2. Improved timing of N application

The timing of fertilizer application varies substantially. Common field practices range from a single treatment unrelated to any specific crop development phase to as many as five split applications carefully timed to maximize yields. Many rice farmers apply N in three split doses — the first dose just before transplanting, the second dose at maximum tillering, and the final

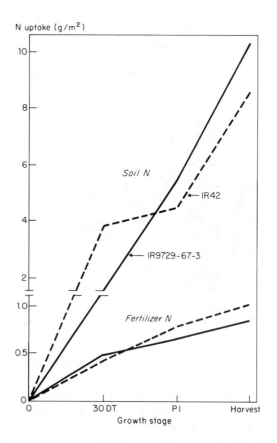

Figure 3. Uptake of soil N and fertilizer N (^{15}N-depleted) in IR42 and an early maturing IR9729-67-3 rice. IRRI, 1982 wet season

dose at or just before panicle initiation. Split application seeks to maintain a pool of labile N while minimizing the risks of massive losses through ammonia volatilization, denitrification, and leaching.

Many farmers in tropical Asia still apply fertilizer directly into the floodwater 1–3 weeks after transplanting despite research that shows the benefits of improved timing and application methods. Broadcasting N fertilizer into floodwater results in extensive N losses to the atmosphere through surface water [4]. For example, in 1984 dry season when 87 kg N/ha as prilled urea (PU) was applied using farmers' timing – one-half N applied into 5 cm of water 10 DT and one-half topdressed into 5 cm of water at 10 days after panicle initiation (PI) – about $18 + 14$ kg N/ha of total N (urea + NH_4^+-N) ended up in floodwater (Figure 4). With researchers' split application – two-thirds broadcast and incorporated without standing water and one-third

Table 1. Grain yield, total N uptake, and fertilizer N uptake by rice crop duration. From Broadbent FE and De Datta SK [1]

Duration (days)	Grain yield (t/ha)	Total N uptake (kg/ha)	Fertilizer N uptake (kg/ha)
		1981 Dry season	
100	4.6	78	17
110	4.5	87	15
120	4.8	97	11
		1981 Wet season	
100	3.6	107	9
110	4.0	121	7
120	3.7	105	6
		1982 Dry season	
100	5.6	74	4
105–115	5.5	90	3
120–135	5.5	91	4

topdressed into 5 cm of standing water at PI — only about $6 + 4$ kg N/ha ended up in floodwater (Figure 4).

Water depth effects on improved timing of N applications. In 1984 dry season when the first dose of applied urea N was broadcast into 5 cm of standing water using researchers' split, about 4 kg NH_4^+-N/ha ended up in floodwater versus only 2 kg NH_4^+-N N/ha when basal N was applied without standing water and incorporated thoroughly into puddled soil (Figure 5). Basal application of urea with 5 cm standing water resulted in 35 kg/ha of total N (urea + NH_4^+-N) in the floodwater versus 8 kg N/ha without standing water (Figure 5). Grain yield was 0.9 t/ha higher with incorporation of a basal dose prior to planting without standing water than with urea application into standing water and incorporation (Table 2). Properly applied researchers' split with basal incorporation without standing water gave similar yield to deep point-placed USG or broadcast and incorporated SCU (Table 2).

These results emphasize the importance of proper water management and basal fertilizer incorporation prior to planting in increasing N use efficiency in lowland rice. This technology is highly relevant in farmers' fields because it appeared to have no risks to the farmers. Based on this and other IRRI research, the Philippine extension program *Masagana-99*, has modified its 16-step fertilizer application recommendations.

3. Deep placement of fertilizer nitrogen

Deep placement of fertilizer in reduced soil has been considered the most efficient method to increase fertilizer N efficiency in lowland rice [12].

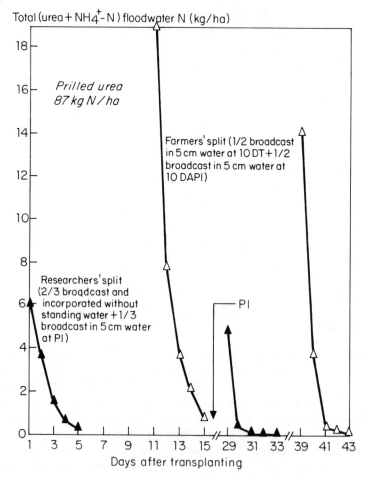

Total (urea + NH$_4^+$-N) floodwater N (kg/ha)

Figure 4. Total (urea + NH$_4^+$-N) floodwater N at 1300 to 1400 hours as affected by timing and method of nitrogen application. IRRI, 1984 dry season. DT = days after transplanting, DAPI = days after panicle initiation, PI = panicle initiation

Deep placement of N fertilizers is becoming increasingly relevant since the introduction of modified urea materials such as urea briquettes, USG, and urea marbles for testing in lowland rice.

Cao et al. [2] reported the most recent results on the effects of placement methods on the fertilizer N recovery and grain yield using ^{15}N-labelled urea in field microplots. USG point placement and uniform placement of PU gave the highest grain yield of 6.4 t/ha in dry season, suggesting that both deep placement methods effectively improve fertilizer N efficiency. The high efficiency (52 kg rough rice/kg applied N) with point placement was due to lower N fertilizer losses as evidenced by the low total N (urea + NH$_4^+$-N) in floodwater [2].

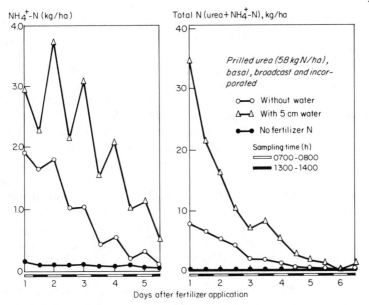

Figure 5. Floodwater NH_4^+-N and total N (urea + NH_4^+-N) at 0700–0800 h and 1300–1400 h after broadcast and incorporation of urea at different water depths. IRRI, 1984 dry season

Table 2. Effects of water depth, urea source, and application method at 87 kg N/ha on the grain yield of IR58[a]. IRRI, 1984 dry season

Urea source	Application method[b]	Water depth during basal fertilizer application (cm)	Grain yield[c] (t/ha)
–	No fertilizer nitrogen	–	2.9 c
SCU	Broadcast and incorporated	0	6.6 a
PU	Researchers' split	0	6.4 a
PU	Researchers' split	5.1	5.5 b
PU	Farmers' split	0	5.4 b
PU	Farmers' split	5.1	5.2 b
PU	Farmers' split	10.0	4.7 b
PU	Farmers' split	14.9	5.1 b
USG	All basal, hand point-placement	5.2	6.6 a

[a] Av of 4 replications. SCU = sulfur-coated urea, PU = prilled urea, USG = urea supergranules, PI = panicle initiation
[b] Researchers' split = 2/3 broadcast and incorporated and 1/3 topdressed with 5 cm standing water at panicle initiation. Farmers' split = equal split-doses at 10 days after transplanting and at 10 days after panicle initiation
[c] Means followed by a common letter are not significantly different at the 5% level by Duncan's multiple range test

Research on USG deep placement under tropical and subtropical conditions was reviewed by Juang [20]. In Taiwan, at the same rate of applied N, USG increased rice yields by an average 20% over PU in a low-yield area such as Taichung but not in a high-yield area such as Pintung.

Results from 16 field experiments conducted by the INPUTS project in 8 Asian countries indicated that USG treatments outyielded PU at the same rate of N in 11 of 16 irrigated sites [20].

Several years of agronomic data from the International Network on Soil Fertility and Fertilizer Evaluation for Rice (INSFFER) network show that deep placement is generally more efficient than the traditional split application of urea [4, 28, 10, 12, 15]. Individual trial results, however, vary by site and season. Tejeda et al. [28] determined the relationships between the efficiency of the different fertilizers and site characteristics, using data from 114 INSFFER trials in 11 countries. Average efficiencies (defined as the increments in rice yield per unit of N applied) up to 56 kg N/ha for USG was found to be 21–24 kg rough rice/kg applied N.

Economic analysis of the same set of trials showed an average return of US $4–7 for every dollar spent on labor for USG application [4]. Recent analysis by Flinn et al. [15] of the 1981 wet season INSFFER trials showed that N applied as USG usually was technically more efficient than N applied as prilled urea, particularly at low N rates. Averaging the data from all experiments, 20–30% less N of the modified N source gave the same yield increment as N applied as PU. Separating the dry season from the wet season INSFFER trials during 1981–82, the average yield in dry season was higher than in wet season at the same N rate (Figure 6). Both USG and sulfur-coated urea (SCU) were equally effective and better than prilled urea during wet season, especially at lower N rates up to 87 kg N/ha. However, at 116 kg N/ha all the 3 forms were similar.

To make deep placement relevant at the farm level, it is critical to develop a series of placement machines that will point-place PU, urea solution, and USG. Several placement machines are currently being developed and evaluated for this purpose at IRRI.

Results from the 1984 dry season showed that hand deep- placed USG produced significantly higher grain yield (average of 58 and 87 kg N/ha) than machine-placed N fertilizer (Table 3). Deep placement machines applied between 46 and 72 kg N/ha for 58 kg N/ha intended rate and between 79 and 91 kg N/ha for 87 kg N/ha intended rate, indicating considerable room for improvement in the accuracy of application rate. Fertilizer N efficiency was greatest with USG hand point placement (Table 3).

4. Controlled release N fertilizers

SCU is the most widely tested coated fertilizers for rice [4, 5, 12]. Studies suggest that slow-release fertilizers such as SCU considerably reduce ammonia

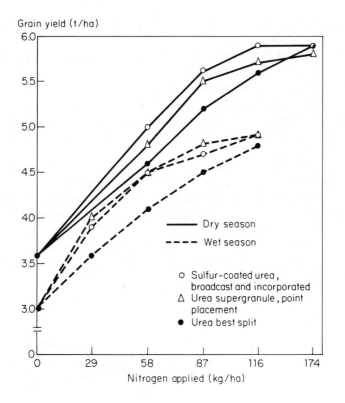

Figure 6. Nitrogen responses of rice with different forms of urea and application rates. Data are averages on 14 dry and 55 wet season trials. Fifth International Trial on Nitrogen Fertilizer Efficiency in Rice, 1981–82 crop seasons. From IRRI [8]

Table 3. Grain yield of IR58 rice as affected by hand- and machine-placed fertilizer N. IRRI, 1984 dry season

Urea source[a]	Application method	Grain yield[b] (t/ha)	Fertilizer efficiency (kg rough rice/kg N)
PU	Band placement by spring auger	3.6 d	26
PU	Band placement by oscillating plunger	3.8 cd	28
USG	Point placement by deep plunger	4.6 b	40
USG	Point placement by press wedge	4.2 bc	28
USG	Point placement by hand	5.2 a	46
PU	Researchers' split	4.4 b	34
PU	Farmers' split	3.4 d	21
–	No fertilizer nitrogen	1.9 e	–

[a] PU = prilled urea, USG = urea supergranules
[b] Av of 4 replications and 2 nitrogen rates. Means followed by a common letter are not significantly different at the 5% level

182

volatilization losses [21] and provide the crop adequate N nutrition through-out the growing season [12].

In the irrigated INSFFER trials SCU effectively increased grain yield in various soils and environments [10]. The average economic response, based on 30% extra cost of SCU over ordinary urea, showed a US $6−7 return for every dollar spent on SCU. Comparative analysis of N sources tested 1981 INSFFER wet season trials revealed that in most cases, N applied as SCU was technically more efficient than N applied as PU [15]. Current INSFFER results confirmed that [18].

In the latest IRRI experiment however, SCU did not increase grain yield over the improved timing, best split N application (Figure 7).

Because of SCU slow-release qualities, N concentration in the soil and floodwater is restricted at any given time, which reduces N losses. Figure 8 shows floodwater NH_4^+-N and total N (urea + NH_4^+-N) were low with SCU and hand deep-placed USG. However, band placement with oscillating plunger of prilled urea resulted in relatively higher NH_4^+-N and total N concentrations in floodwater (Figure 8).

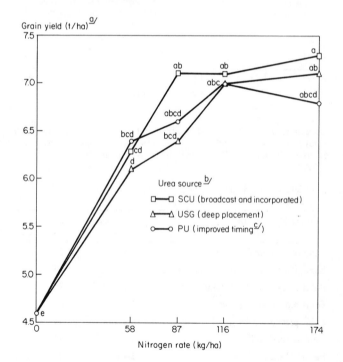

Figure 7. Effect of forms, rates, and methods of urea application on grain yield of irrigated IR58 rice. IRRI, 1984 dry season. [a]Av of 4 replications. Separation of means by Duncan's multiple range test at the 5% level. [b]SCU = sulfur-coated urea, USG = urea supergranule, PU = prilled urea. [c]Two-thirds broadcast and incorporated at planting; one-third topdressed at 5−7 days before panicle initiation

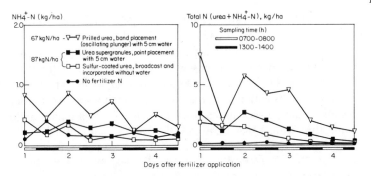

Figure 8. Floodwater NH_4^+-N and total N (urea + NH_4^+) at 0700–0800 h and 1300–1400 h as affected by sources of urea, application methods, and water depth. IRRI, 1984 dry season

Based on those results and the results summarized by Savant and De Datta [26], the benefits of using SCU to improve the N utilization efficiency in lowland rice cannot be over emphasized. The benefits of SCU are:

1. increased fertilizer use efficiency,
2. reduced N losses,
3. reduced toxicity to rice seedlings,
4. reduced application frequency, and
5. fulfillment of S requirements [8].

Research to identify alternate slow, and controlled-release N fertilizers is in progress.

5. Use of nitrification and urease inhibitors

Among N fertilizer losses, nitrification and denitrification have traditionally been considered the main loss mechanisms. Recent results suggest that NH_3 volatilization losses are most important if partial pressure of NH_3 is high in floodwater, wind speed is high, and the rice crop is too young to absorb faster than the loss mechanisms [14].

On the other hand, in studies conducted in New South Wales, Australia on a clay soil with pH 8.2 Simpson et al. [27] reported 46% of applied N was lost from the water-soil-plant system during the first 11 days after application. Only 11% was volatilized as ammonia, despite floodwater pH values up to pH 10 and strong winds. Gaseous loss as nitrous oxide was negligible, and similar to earlier results [4]. Leaching of N beyond 100 cm soil depth was also negligible.

Results from Simpson et al. [27] indicate 35% of the applied N was lost through denitrification of nitrite and nitrate after nitrification of ammonium near the soil surface. Final fertilizer N recovery by the crop was only 17%. These results suggest remaining challenge to develop management practices and suitable nitrification inhibitors to minimize denitrification losses and improve N use efficiency in lowland rice.

Nitrification inhibitors. Early results of experiments with nitrification inhibitors were reported by Prasad et al. [25]. IRRI results show that application of PU with 10 or 15% dicyandiamide (DCD) during the final harrowing produced yields comparable to those with split-applied PU without DCD in lowland IR36 rice [11]. DCD breakdown is high with high temperature in the tropics. Results with nitrapyrin have been inconsistent in the tropics.

Research should be accelerated to identify nitrification inhibitors for the high temperature tropics that are better than DCD and nitrapyrin.

Urease inhibitors. De Datta and Craswell [9] suggested that research on urease inhibitors should be conducted. Urease inhibitors can reduce ammonia volatilization losses from broadcast urea applications by delaying, and then reducing, the buildup of ammonium concentration in the floodwater [30]. Field results at IRRI [12] and Maligaya [13] confirmed the delaying, reducing action. Although phenyl phosphorodiamidate (PPD) retarded urea hydrolysis, it did not increase grain yield and total N uptake compared with a control without urease inhibitors [11]. In a recent urease inhibitor study, floodwater pH was unaffected and displayed a normal diurnal pattern. Contrary to earlier findings, applying PPD with urea significantly increased N uptake and grain yield in an IRRI experiment where urea or urea + 1% PPD were topdressed 26 DT. Grain yield was not increased by applying PPD in a similar experiment at the Maligaya Rice Research and Training Center in Nueva Ecija, Philippines where urea + 1% PPD were applied 18 DT [19]. Research should be carried out to identify specific factor(s) that cause PPD to give variable results in different sites.

6. Complementary practices to improve N use efficiency

Several complementary management practices can markedly increase the N fertilizer efficiency. They are:

1. planting at optimum time to expose the rice crop to maximum solar energy at the reproductive and ripening periods,
2. close planting and optimum geometry to maximize panicle production per unit area,
3. early, thorough weed removal to maximize utilization of water and nutrients by the crop,
4. repairing of levees to minimize seepage,
5. thorough puddling and leveling of fields to conserve water and nutrients,
6. use of straw, azolla, and organic and green manures as supplementary N sources, and
7. other good crop husbandry practices to minimize crop damage to pests and optimize growth for maximum nutrient use efficiency.

These agronomic practices are most important to follow under resource-scarce situations, as in most tropical Asian countries, than under resource-abundant situations that exist in many temperate countries.

In the present fertilizer use-intensive rice production technology, agronomists and soil scientists should pay greater attention to management of the natural supply of plant nutrients [23]. Natural nutrient sources for a rice soil are mainly edaphic constituents, rainwater and irrigation water, the rice cultural system per se, rice based cropping systems, recycling of crop residues and other organic waste materials and biological N fixation.

Deep tillage and good land preparation promote full and efficient utilization of the native soil nutrients and enhance expensive production inputs such as N fertilizers.

It is apparent that maximum average 4 to 5 t/ha rice yields can be obtained with deep tillage, high fertilizer rates, good irrigation, use of input-responsive modern rices and optimum crop husbandry practices. The research challenge for increased tropical Asian rice production is not only to maximize rice yields but also to maximize input use efficiency, particularly N use efficiency, in lowland rice.

References

1. Broadbent FE and De Datta SK (1984) Differences in response to nitrogen fertilizer by rice varieties/lines. Agronomy Abstract. 1984 Annual Meetings, American Soc Agron, Las Vegas, Nevada, USA
2. Cao Zhi-Hong, De Datta SK and Fillery IRP (1984) Effect of placement methods on floodwater properties and recovery of applied nitrogen (^{15}N-labeled urea) in wetland rice. Soil Sci Soc Am J 48:197–203
3. Craswell ET and Vlek PLG (1979) Fate of fertilizer nitrogen applied to wetland rice. In International Rice Research Institute. Nitrogen and rice, 175–192, Los Baños, Laguna, Philippines
4. Craswell ET and De Datta SK (1980) Recent developments in research on nitrogen fertilizers for rice. IRRI Research Paper Series 49, 11 pp
5. Craswell ET and Vlek PLG (1982) Nitrogen management for submerged rice soils. In Symposia Papers. II. Vertisols and rice soils of the tropics. Transactions 12th Cong Int Soil Sci 3:158–181, 8–16 February 1982, New Delhi
6. De Datta SK (1981) Principles and practices of rice production. John Wiley & Sons, New York, 618 pp
7. De Datta SK, Garcia FV, Chatterjee AK, Abilay WP Jr, Alcantara JM, Cia BS and Jereza HC (1979) Biological constraints to farmers' rice yields in three Philippine provinces. IRRI Paper Series 30, 69 pp
8. De Datta SK, Craswell ET, Fillery IRP, Calabio JC, and Garcia FV (1981) Alternative strategies for increasing fertilizer efficiency in wetland rice soils. Paper presented at the International Rice Research Conference, 27 April–1 May 1981. International Rice Research Institute, Los Baños, Laguna, Philippines
9. De Datta SK, and Craswell ET (1982) Nitrogen fertility and fertilizer management in wetland rice soils. In International Rice Research Institute. Rice research strategies for the future 283–316
10. De Datta SK, and Gomez KA (1981) Interpretive analysis of the international trials on nitrogen fertilizer efficiency in wetland rice. In Fertilizer International 1–5, The British Sulphur Corp. Ltd., Parnell House, England
11. De Datta SK (1985) Availability and management of nitrogen in lowland rice in relation to soil characteristics. In International Rice Research Institute. Wetland soils, characterization, classification, and utilization, 247–267, Los Baños, Laguna, Philippines
12. De Datta SK, Fillery IRP and Craswell ET (1983) Results from recent studies on nitrogen fertilizer efficiency in wetland rice. Outlook on Agriculture 12:125–134

13. Fillery IRP and De Datta SK (1985) Effect of N sources and a urease inhibitor on NH_3 loss from flooded rice fields (i) Ammonia fluxes and ^{15}N loss. Soil Sci Soc Am J (in press)

14. Fillery IRP, Simpson JR and De Datta SK (1985) Ammonia volatilization and ^{15}N loss in wetland rice yields receiving different fertilizer N management. Fertilizer Research (in press)

15. Flinn JC, Velasco LE and Kaiser K (1982) Comparative efficiency of N-sources, 5th wet season INSFFER trials on irrigated rice. Paper presented at the International Rice Research Conference, 19–23 April 1982. International Rice Research Institute, Los Baños, Philippines

16. FAO (Food and Agriculture Organization of the United Nations) (1983) Ceres 96. FAO review on agriculture and development, 16(6):11

17. Herdt RW (1981) Focusing research on future constraints to rice productions. Paper presented at the International Rice Research Conference, 27 April– 1 May, 1981, International Rice Research Institute, Los Baños, Laguna, Philippines

18. IRRI (International Rice Research Institute) (1983) Preliminary report on the fifth international trial on nitrogen fertilizer efficiency in rice (1981–82), International Network on Soil Fertility and Fertilizer Evaluation for Rice, Los Baños, Laguna, Philippines

19. IRRI (International Rice Research Institute) (1984) Research highlights for 1983, Los Baños, Philippines, 121 pp

20. Juang TC (1980) Increasing nitrogen efficiency through deep placement of urea supergranules under tropical and subtropical paddy conditions. In Asian and Pacific Council, Food and Technology Center, Increasing Nitrogen Efficiency for Rice Cultivation 83–101, Taipeh, Taiwan, Republic of China

21. Mikkelsen DS, De Datta SK and Obcemea WN (1978) Ammonia volatilization losses from flooded rice soils. Soil Sci Soc Am J 42:725–730

22. Murayama N (1979) The importance of nitrogen for rice production. In International Rice Research Institute, Nitrogen and rice 5–23, Los Baños, Laguna, Philippines

23. Patnaik S and Rao MV (1979) Sources of nitrogen for rice production. In International Rice Research Institute Nitrogen and rice, 25–44, Los Baños, Laguna, Philippines

24. Prasad R and De Datta SK (1979) Increasing fertilizer nitrogen efficiency in wetland rice. In International Rice Research Institute Nitrogen and rice 465–484, Los Baños, Laguna, Philippines

25. Prasad R, Rajale GB and Lakhdive BA (1971) Nitrification retardants and slow release nitrogen fertilizers. Adv Agron 23:337–383

26. Savant NK and De Datta SK (1982) Nitrogen transformation in wetland rice soils. Adv Agron 35:241–302

27. Simpson JR, Freney JR, Wetselaar R, Muirhead WA, Leaning R and Denmead OT (1984) Transformations and losses of urea nitrogen after application to flooded rice. Aust J Agric Res 35:189–200

28. Tejeda HR, Craswell ET and De Datta SK (1980) Site factors affecting the efficiency of nitrogen fertilizers in INSFFER experiments. Paper presented at the INSFFER meeting, 24 April 1980, International Rice Research Institute, Los Baños, Philippines

29. Vega MR (1982) Low-input technology for optimum productivity of rice. Presented at the 13th Annual Meeting of the Crop Sci Soc of the Philippines, Cebu City, April 28–30, 1982

30. Vlek PLG, Stumpe JM and Byrnes BH (1980) Urease activity and inhibitors in flooded soil systems. Fert Res 1:191–202

31. Wortman S and Cummings RW Jr (1978) To feed this world, the challenge and strategy. The Johns Hopkins Press, Baltimore, Maryland, 440 pp

32. Yoshida S (1981) Fundamentals of rice crop science. International Rice Research Institute, Los Baños, Laguna, Philippines, 269 pp